Migration in the Western Balkans

Migration in the Western Balkans rectifies the under-investigation by migration scholars of the Western Balkans region, by bringing together recent research at a time when migration is a hot topic for the future of Europe.

The book explores issues such as the complex geopolitics of the region, the relationship between migration and development, diasporas and refugees and humanitarianism. Expert contributors present new research on economic migration, forced migration, diaspora formation and return migration at a time when migration is of crucial relevance for the future of Europe. The chapters shed new light on the multiple migration dynamics of a region which has had a troubled past yet stands on the threshold of EU membership.

As a theatre of multiple migration processes *Migration in the Western Balkans* reveals new information on the region, and will be of great interest to scholars of migration, the Balkans and geopolitics.

The chapters were originally published as a special issue of the *Journal of Balkan and Near Eastern Studies*.

Russell King is Professor of Geography at the University of Sussex, UK, and Visiting Professor in Migration Studies at Malmö University, Sweden. He is President of the Western Balkans Migration Network (WB-MIGNET).

Nermin Oruc is Director of the Centre for Development Evaluation and Social Science Research (CDESS), Sarajevo, Bosnia and Herzegovina, and Visiting Atlantic Fellow at the International Inequalities Network at London School of Economics and Political Science, UK. He is Chair of WB-MIGNET.

Migration in the Western Balkans

What do we know?

Edited by
Russell King and Nermin Oruc

Routledge
Taylor & Francis Group
LONDON AND NEW YORK

First published 2020
by Routledge
2 Park Square, Milton Park, Abingdon, Oxon, OX14 4RN

and by Routledge
52 Vanderbilt Avenue, New York, NY 10017

Routledge is an imprint of the Taylor & Francis Group, an informa business

British Library Cataloguing-in-Publication Data
A catalogue record for this book is available from the British Library

ISBN13: 978-0-367-42512-8

Typeset in Minion Pro
by codeMantra

Publisher's Note
The publisher accepts responsibility for any inconsistencies that may have arisen during the conversion of this book from journal articles to book chapters, namely the inclusion of journal terminology.

Contents

Citation Information

The chapters in this book were originally published in the *Journal of Balkan and Near Eastern Studies*, volume 21, issue 1 (February 2019). When citing this material, please use the original page numbering for each article, as follows:

Chapter 1
Migration in the Western Balkans – Trends and Challenges
Russell King and Nermin Oruc
Journal of Balkan and Near Eastern Studies, volume 21, issue 1 (February 2019), pp. 1–10.

Chapter 2
From Source to Corridor: Changing Geopolitical Narratives about Migration and EU-Western Balkans Relations
Ferruccio Pastore
Journal of Balkan and Near Eastern Studies, volume 21, issue 1 (February 2019), pp. 11–26.

Chapter 3
Balkans and Baltics: On Migration as a Factor of Regional Peripheralization
Aija Lulle
Journal of Balkan and Near Eastern Studies, volume 21, issue 1 (February 2019), pp. 27–42.

Chapter 4
Refugees at the Gates of the EU: Civic Initiatives and Grassroots Responses to the Refugee Crisis along the Western Balkans Route
Chiara Milan
Journal of Balkan and Near Eastern Studies, volume 21, issue 1 (February 2019), pp. 43–60.

Chapter 5
The Floating Pumpkin Syndrome: Forced Migration, Humanitarian Aid, and the Culture of Learned Helplessness
Nina Bosankić, Enisa Mešić and Bojan Šošić
Journal of Balkan and Near Eastern Studies, volume 21, issue 1 (February 2019), pp. 61–73.

Chapter 6
The Serbian Diaspora's Political Views: A Study on the 2017 Serbian Presidential Election
Tena Prelec
Journal of Balkan and Near Eastern Studies, volume 21, issue 1 (February 2019), pp. 74–89.

Chapter 7

The Effects of Remittances on Education in a Post-Conflict Society: Evidence from Bosnia-Herzegovina
Nermin Oruc, Ian Jackson and Geoffrey Pugh
Journal of Balkan and Near Eastern Studies, volume 21, issue 1 (February 2019), pp. 90–103.

Chapter 8

The Greek Debt Crisis and Albanian Return Migration
Kalie Kerpaci and Martin Kuka
Journal of Balkan and Near Eastern Studies, volume 21, issue 1 (February 2019), pp. 104–119.

Contributors

Nina Bosankić is Assistant Professor in the Psychology Programme at the Faculty of Arts and Social Sciences at the International University of Sarajevo, Bosnia and Herzegovina.

Ian Jackson is Senior Lecturer in Economics and Finance at Wolverhampton University Business School, UK.

Kalie Kerpaci is Lecturer in Psychology at Aleksander Moisiu University, Durrës, Albania.

Russell King is Professor of Geography at the University of Sussex, UK, and Visiting Professor in Migration Studies at Malmö University, Sweden. He is President of the Western Balkans Migration Network (WB-MIGNET).

Martin Kuka is Lecturer in Economics at Aleksander Moisiu University, Durrës, Albania.

Aija Lulle is a migration scholar and Lecturer in Human Geography at Loughborough University, UK. She has previously worked at the Universities of Latvia, Sussex (UK) and Eastern Finland.

Enisa Mešić is a psychologist and a PhD candidate at the Department of Economics at Sarajevo School of Science and Technology in Sarajevo, Bosnia and Herzegovina.

Chiara Milan is Marie Skłodowska-Curie Fellow at the Centre for Southeast European Studies of the University of Graz, Austria. She holds a PhD in Political and Social Sciences from the European University Institute, Florence, Italy.

Nermin Oruc is Director of the Centre for Development Evaluation and Social Science Research (CDESS), Sarajevo, Bosnia and Herzegovina, and Visiting Atlantic Fellow at the International Inequalities Network at London School of Economics and Political Science, UK. He is Chair of WB-MIGNET.

Ferruccio Pastore received his PhD from the European University Institute, Florence, Italy. Currently, he is Director of FIERI (Forum Internazionale ed Europeo di Ricerche sull'Immigrazione) in Turin, Italy.

Tena Prelec is completing her PhD in the School of Law, Politics and Sociology at the University of Sussex, UK. She is also a Research Associate at LSEE-Research on South Eastern Europe, European Institute at London School of Economics and Political Science, UK.

Geoffrey Pugh is Professor of Economics at Staffordshire University Business School, Stoke-on-Trent, UK.

Bojan Šošić serves as a member of the Board for Psychiatric and Neurological Research in the Department of Medical Sciences at the Academy of Sciences and Arts of Bosnia and Herzegovina.

Migration in the Western Balkans – Trends and Challenges

Russell King and Nermin Oruc

ABSTRACT

This introductory paper sets the scene for the special issue on migration in the Western Balkan region. First, we briefly outline the background to the research networking initiative in the Western Balkans—'WB-MIGNET'—which resulted in producing the set of papers presented here. Then we describe the context in which the analysis of migration, return and development in the Western Balkans takes place, highlighting the role of diasporas and return migration. The region exemplifies a wide range and interrelation of migratory forms, including temporary and permanent labour migration, forced migration of refugees, temporary displacement, high-skilled migration and transit migration. Quite apart from the scale and diversity of migratory phenomena in this region, the special relevance of the Western Balkans rests on mass emigration, predominantly of educated young people, and their possible return, including to post-conflict areas, and role in the development of their home countries. Such processes within the Western Balkans also offer useful lessons for designing future migration policies in Europe and worldwide. The final section of this introductory paper summarizes the papers that follow and highlights their originality and key findings.

The Western Balkans and migration

The Western Balkans have a particular relevance in the study of migration in its various forms, including labour migration, high-skilled migration, forced migration, diaspora formation and return migration—all of which are present in this region. Migration has also shot up the European Union policy agenda, due to both the legacy of past migrations, which were intrinsic to the formation of the Common Market, European Community and then the EU, and more particularly to the emergence of new challenges—often labelled migration and refugee 'crises'—in the last few years.[1] Within the Western Balkan region, countries such as Albania, Bosnia-Herzegovina (BiH) and Kosovo experienced mass displacements of their populations during the early transition periods coincident with the end of the communist/socialist era and the break-up of Yugoslavia. Ethno-national conflict and prolonged economic crisis have been, often in tandem, the main drivers of exodus from the region. And yet this region as a whole remains under-investigated by migration scholars.[2] This special issue represents an effort to reduce this gap in policy-relevant research.

This issue is one of the first concrete outputs of an initiative called the Western Balkans Migration Network (WB-MIGNET), which gathers migration scholars from within the region and connects them both with each other and with experts from outside who are active in the region. The countries currently included as core members of the Network are Albania, Bosnia and Herzegovina, Croatia, Kosovo, Macedonia, Montenegro and Serbia. The activities of the Network include the organization of conferences and other events, linkage with other networks and drafting new projects, the outcome of which is research papers such as those contained in this issue. The papers that follow represent a selection of those presented at the Network's second annual conference, held in Sarajevo in May 2017.[3] They cover important topics such as the geopolitics of migration in the region, forced migration and return, the role of diaspora in political life, recent refugee flows through the region, and comparison with other regions.

Trends and challenges

Emigration from less to more developed countries has gathered pace in the last two decades. According to the International Organization for Migration, there were globally 244 million migrants in 2015 (3.3% of the world's population), a significant increase over the 155 million estimated for 2000 (2.8% of global population).[4] Whilst most of this movement of people is driven by economic opportunities in the countries of immigration, recent years have witnessed an increased number of conflicts, which also 'produce' migration in the form of forced displacement and resettlement elsewhere. Meanwhile, return migration is becoming ever more salient in recent years, as the current worldwide economic recession impacts on the livelihoods of migrants in various host countries (notably in Greece and other countries of Southern Europe) and pushes them to return to their home countries or to move on elsewhere (see the paper by Kerpaci and Kuka in this special issue). Moreover, the new wave of literature on the consequences of migration generally emphasizes its positive impact on home countries through remittances, diaspora activities and return migration.[5] Most of the existing research on the economic ties between migrants and their countries of origin has focused on remittance flows, and remittances are certainly important for the economic survival of many households in the poorer countries of the Western Balkan region.[6] More recently this has expanded to include the potentially transformative political and social roles that migrants and diaspora members can play in their countries of origin, through such varied forms of activity as political and party support, media and cultural events, and civil society development.[7] Less attention has been paid to diaspora members' investments in countries of origin.[8]

Like lower-skilled labour migrants, the emigration of high-skilled individuals has a mix of positive and negative effects for the source country. First, and again like their less-skilled compatriots, high-skilled migrants often send remittances back home, unless they are from wealthy families. These remittances have the potential to have multiple impacts on the members of the household who remain in the home country, one of the most significant being investment in education of younger family members (see the paper in this special issue by Oruc et al.). At a more macro scale, diaspora involvement has a range of other positive effects on a country's economic development, through its facilitation of trade, foreign direct investment, tourism and other relations with the migrant-origin country. High-skilled return migration brings an important channel of transfer of financial savings, human capital, cultural and other assets from the migrants' host to their home countries. These gains may overcome the principal negative effect of highly skilled migration—brain drain. The magnitude of the positive effects that diaspora involvement and return migration can bring to the source economy depends on the scale of the impact, its evolution over time, and the involvement of national and local governments, in coordination with diaspora and migrant organizations, in maximizing the benefits.

Return migration to areas which were previously consumed by conflict deserves a separate analysis, for several reasons. First, conflict-induced migration is predominantly family migration, since the migration decision is 'forced' and not shaped by the need to seek employment or to diversify the risks of household economic hardship. This may result in a weaker incentive to return. Second, the pull factors of return migration are different in the case of conflict-induced migration. Rather than the possible pulls of economic improvement, nostalgia for the home community, or the wish to rejoin family members 'left behind', conflict migrants usually harbour negative experiences and even traumatic memories of the situation that drove them away. Also, the situation in their home area might not be sufficiently attractive, economically or in terms of inter-ethnic security, for them to decide to return.[9] These experiences of various parts of the Western Balkans that went through these cycles of mass departures and returns may be of relevance for current and future policies of return as regards recent refugees to the EU.

Specific country experiences

Albania experienced a dramatic period of mass emigration starting in 1991, following the fall of the communist regime that had kept the country closed off from the outside world since the end of World War II. Tens of thousands crossed the mountain border into Greece or took to boats to sail across to southern Italy.[10] Another large-scale exodus took place in 1997 as a result of the (inevitable) collapse of corrupt pyramid savings schemes, which disastrously affected a large share of the Albanian population. With emigration continuing in subsequent decades, it is currently estimated that Albania has 'lost' between a quarter and a third of its total population to emigration.[11]

During the war in 1991–1995, around a quarter of the total population of BiH left the country due to the 'ethnic conflict', mainly fleeing to countries in Western Europe. Whilst on the one hand much of this was whole-family migration inclusive of all socio-economic groups, there was an over-representation of highly skilled individuals. Uvalic estimated that around 85% of PhD degree holders and 75% of Master's degree holders emigrated during the wars.[12] Even more dramatic, in relative terms, was the near-total displacement of the Kosovo population by the ethnic persecution of the brutal Milosevic regime in the late 1990s.

Emigration from other countries during the 1990s was also considerable, due to the war (Croatia), economic sanctions against the Milosevic regime in Serbia, and economic crisis in Macedonia. All countries of the region, due to ongoing political instability and protracted economic crisis, have experienced continued emigration in the twenty-first century. The available evidence from each country suggests that migrants are positively self-selected: i.e. the share of highly skilled individuals among migrants is usually considerably larger than their share in the total population.

Return migration to the Western Balkans has its specific characteristics and is diverse in form. Return to Albania is a relatively recent phenomenon, driven more by economic crisis in Greece than by strong pull factors from the Albanian economy, which remains weak and productively underdeveloped.[13] The situation with the return of refugees has been different. Most of the refugees from BiH in the early 1990s received 'temporary protection' status in EU countries, which then allowed host countries to 'repatriate' those people when the conflict was over and it was deemed 'safe' to return. This happened already at the end of 1995 and so during 1996–1998 most of those under temporary protection were sent back to Bosnia.[14] The mandatory repatriation process from Germany started immediately after the cessation of hostilities in Bosnia, which was 'questionable on moral and humanitarian grounds, given current circumstance in BiH'.[15] Moreover, most of the ex-refugees were returned to a place different from the one they lived in before the war, due to the re-drawing of the ethno-religious map by opposing warring parties. Thus, they became internally displaced instead of refugees. The return of refugees to Kosovo was broadly similar, with the major difference that their stay abroad was a few months and not four years as in the case of BiH. Soon after the air strikes of the NATO forces and the withdrawal of the Serbian army from Kosovo, the refugees returned home. Return migration to Croatia, Montenegro and Serbia, where voluntary migration has predominated, is largely a return of retired workers after a long stay abroad. Such a return poses challenges of reintegration, health and service provision, and portability of social benefits.

The large scale of emigration from the region, primarily before and during the 1990s, has created large diaspora communities for each country. Because of the generational effect, including intermarriage, estimates of the size of diasporas are approximate and often exaggerated. But two examples of more recently formed diasporas are likely to be fairly accurate: both Albania (1.4 million diaspora Members) and BiH (2 million) have diaspora populations equivalent to half the population living in the respective countries.[16] The current size of the Serbian diaspora is around 1 million, which is equivalent to 15% of its population in the country.

Thus far, these large migrant and diaspora populations have mainly been contributing to the economies of their home countries by sending remittances. This has created a strong and potentially dangerous reliance of these countries on these financial inflows, which in most cases greatly outstrips the scale of foreign direct investment. Only recently have the countries of the region recognised that contributions from the diaspora can go beyond remittances and include investment in businesses, knowledge transfer, tourism and cultural contributions. There is a move to put policies in place to encourage this more diversified economic and cultural relationship. Still, the effectiveness of such government actions, and their durability, remain uncertain. Recent evidence also suggests that a wider involvement of the diaspora includes fomenting their interest in more active political participation.[17] Such a choice was largely driven by the motivation of nationalist governments to secure more votes, in the belief that migrant and diaspora populations would support the nationalist agenda. But more recent migrants are often in opposition to the current political establishments in their home countries, which in some cases may induce a different behaviour of governments towards their migrant diasporas (see the paper by Prelec in this issue).

The papers: highlights and key findings

Seven papers follow this editorial introduction. The first three treat the region as a whole, from a variety of angles, whilst the subsequent four papers focus on migration-related processes in individual countries—Bosnia-Herzegovina (two papers, on very different topics), Serbia and Albania.

The first paper, by Ferruccio Pastore, analyses the changing role of the 'migration factor' in relations between the EU and the Western Balkans. Through a close examination of 'geopolitical narratives', Pastore shows that the nature of the 'factor' of migration changed through three phases: (i) framing the region as a reservoir of labour migration (in Pastore's words, 'Gastarbeiter-providers') for West European, especially West German, industry in the 1960s and 1970s; (ii) a situation of irregular migration, population displacement and refugee flows associated with the break-up of Yugoslavia and the Albanian emergency in the early 1990s; and (iii) finally to a narrative imbued with 'crisis' tropes linked to the functioning of the region in 2015–2016 as a tourist corridor or 'Western Balkan route' for refugees and others fleeing the Syrian civil war and other conflicts in the Near East. Throughout his analysis, Pastore is attentive to the dangers of 'geopolitical determinism' and narrative simplification based on a simplistic and essentialized understanding of the relations between migration and territories. Pastore disaggregates the crudely simplified 'Balkan route' into a number of changing itineraries through and beyond the region, and fundamentally questions the tripartite 'migration functions; of territories which are routinely categorized as 'sending', 'receiving' or 'transit' spaces. This simplistic classification 'can generate a geopolitical trap by condemning that particular country or region to a marginal position in the global geography of human mobility, with clear prejudice to its developmental prospects'.

The next paper, by Aija Lulle, compares the Western Balkans with the three Baltic states (Estonia, Latvia and Lithuania) and discusses possible lessons to be learnt from such an inter-regional comparison between these clusters of small states at opposite ends of Europe. In sustaining this comparison, Lulle develops an argument based on her analysis of scientific and policy discourses relating to 'peripherality' and 'peripheralization' in Europe. She starts from the notion that regions such as the Baltics or Western Balkans are certainly geographical realities, but they are also, more importantly, socially constructed.[18] This social construction takes place according to specific or nominated criteria, such as their migration status as regions of emigration or of immigration. Both the Baltic and the Western Balkan regions have suffered peripheralization as a result of the fall-out from the end of socialist rule and the dismemberment of the Soviet Union and Yugoslavia. The result, in both cases, has been mass emigration, brain drain, depopulation and declining birth rates, but this runs the risk of being an exaggerated stereotype. Her empirical analysis, based on a close reading of 54 research papers on the two regions, focuses on four themes: (i) East–West emigration,

(ii) depopulation and brain drain, (iii) transit migration and irregular border-crossing (mainly for the Western Balkans), and (iv) weak development, discouraging both return migration and immigration. All these discursive themes work towards a broader discourse of peripheralization—the social construction of both regions as dominated by economic and to some extent political fragility and wholesale emigration. Estonia stands somewhat apart as 'semi-periphery'. Because migration is dynamic—it swells and shrinks, and net emigration can transition to net immigration—there is an obvious danger of 'fixing' the peripheral status of regions like the two under consideration, denying them a more optimistic future.

The Western Balkans has a long history of dealing with refugees. However, the so-called 'refugee crisis' that started in early 2015 grew rapidly into an unprecedented phenomenon in its scale, external causes and dynamically evolving nature. The paper by Chiara Milan analyses the complex issues surrounding the refugee flows arriving and passing through the Western Balkan region in 2015–2016. Her main focus is on the civic response to the refugee emergency in two West Balkan states—Macedonia, the entry-point; and Serbia, the bottleneck. Based on 15 in-depth interviews with key-informant activists, Milan explores the various solidarity and humanitarian initiatives organized in the region by groups of the local population to ease the plight of the refugees in transit or stuck at border crossings and holding centres. The paper also examines the extent to which new alliances were formed amongst non-state actors in a region traditionally characterized by deep national divisions and ethno-political enmities and by a low level of grassroots activism. In this respect, the refugee crisis represented a kind of watershed at the societal level, for it spurred a civic response which rose above the divisive behaviour of the various states and political leaders. Milan's original take on the phenomenon of civic response to the emergency situation is also grounded in her recognition of the importance of the emotional and moral involvement of local people, which was often rooted in their own experience of uprooting and displacement.

The refugees theme is continued in the fourth paper, co-authored by Nina Bosankić, Enisa Mešić and Bojan Šošić. Using the metaphor of a 'floating pumpkin'—that bobs about on the water, following its streams and currents yet remains intact with its own integrity—the authors trace the troubled trajectories of returnees and internally displaced persons (IDPs) in Bosnia and Herzegovina. In the post-conflict era, returnees and IDPs are caught between their dependence on humanitarian aid and the difficulties of securing employment and a sustainable income in their still troubled country, with its ongoing economic political and ethnic problems. A telling statistic is that BiH has the highest rate of youth unemployment in the world: 63%. Based on interviews, questionnaires and focus-group discussions with more than 2000 people, collected over a 10-year period, Bosankić et al. counterpose two notions. On the one hand they document a culture of 'learned dependency' which is, largely unintentionally, imposed on returnees and IDPs by donor agencies that want to help and support people, but actually inculcate, through a too-authoritarian stance, a feeling of helplessness. On the other hand, they also find evidence of the opposite effect, namely that adversity awakens certain skills and qualities in people that were dormant during the conflict years. Drawing on social psychology, the authors suggest that processes analogous to 'post-traumatic growth' might be increasing returnees' potential to develop practical survival instincts and a kind of 'personal wisdom' about the challenging world around them.

Tena Prelec analyses the Serbian diaspora's political views in a study on the 2017 Serbian presidential election. Prelec first addresses the problems, already noted earlier in this introduction, of enumerating the size of a diaspora: in the Serbian case these estimates range from less than 1 million according to the International Organization for Migration up to 3 million in unofficial estimates. Either way, the 9523 foreign-resident Serbs who voted in the 2017 presidential election are a small number, and one of the author's key findings was the logistical difficulties faced by Serbian citizens abroad in registering their right to vote, and then actually casting it. The main thrust of Prelec's paper, however, is an interrogation of the established assumption that migrants and diasporas hold more extreme—often nationalistic—views than their peers who live in the home country.[19] Prelec's contribution, based on two surveys ($N = 322$) of the Serbian diaspora taken immediately before, and then soon after, the April 2017 election, generates a very different set of insights. First, there was found to be stark majority opposition to the incumbent government, a rejection of the nationalist path that the country has taken, and widespread support for more 'liberal' opposition candidates. Second, the key issues that preoccupied the diaspora voters were not the supposedly 'big' geopolitical issues of Kosovo, EU membership and relations with Russia, but more general concerns with governance, corruption, the economy, the labour market and education. The question that remains, however, concerns the strategy of the ruling elite: should it work to extend the enfranchisement of the diaspora if there is a majority of progressively minded diasporans who would vote for opposition parties and candidates?

In the penultimate paper Nermin Oruc, Ian Jackson and Geoffrey Pugh analyse the effects of remittances on receiving households in Bosnia and Herzegovina, with special reference to their utilization on rates of school enrolment. This can be regarded as a novel take on the topic of migration and 'brain gain': improving the quality of brains at home through the investment of remittances in education at secondary and tertiary level, instead of encouraging the 'emigrant brains' to return. The paper departs from the fact that 'ethnic cleansing' in BiH resulted in an emigration that was exogenously determined, which allows the specification of a regression model without assuming endogeneity or worrying about reverse causation and therefore worrying about the methodological challenge of correctly choosing the instrumental variables. The paper also introduces a new approach to the estimation by measuring the effect (investment in children's education) along the income variable, using income distribution by quintiles. The evidence presented in the paper, drawn from the 'Living in BiH' survey, conducted by the BiH Office of Statistics in 2004 (valid observations 1367), shows that remittances contribute to the relaxation of the 'liquidity constraint' on investment in education only in the upper quintiles of the income distribution, above the poverty line. Presumably for the lower quintiles, remittances are prioritized for spending on 'basic needs', or perhaps education is a lesser priority anyway.

Finally, Kalie Kerpaci and Martin Kuka analyse the relatively recent phenomenon of Albanian migrants returning to Albania in the wake of the Greek financial crisis, which first became a potential return migration 'push factor' in 2009 but is still ongoing.[20] Albania stands somewhat apart in the Western Balkans, an outlier in the sense that it was not part of the former Yugoslavia, but also distinctive within the entire former 'Eastern bloc' for its rigid communist orthodoxy, extreme isolationism and ideology of economic self-sufficiency. Unlike the former Yugoslavia, which opened its doors to emigration already in the 1960s, Albania's borders remained sealed until 1990–1991. This means that it offers itself nowadays as a rather unique setting for post-1990 *ex novo* emigration, the scale and suddenness of which took both the country itself and its Western neighbours (chiefly Greece and Italy) by surprise.[21] As Albania's contiguous EU neighbour, Greece has played host to the largest exodus of Albanians, who found work in a variety of low-skill niches in the Greek economy. The faltering and near-collapse of that economy over the past decade has rendered life for many Albanian migrants there extremely precarious. Drawing on 25 in-depth interviews with returnees in Tirana, Kerpaci and Kuka analyse the reasons that, in many cases, more or less forced them to depart from Greece, the obstacles they faced in their reintegration into the Albanian society and economy, and the way these barriers shape their future intentions. As expected, return was reluctantly engaged in, since migrants had put down roots in Greece, but their decision was to a large extent forced by the effects of the Greek crisis—falling incomes, unemployment, exhaustion of savings, failure of a business—against which prospects in Albania, especially in Tirana, actually seemed brighter. However, the rapid changes that have taken place in Tirana in their absence proved disorientating for some returnees, who struggled with what Kerpaci and Kuka call Albania's 'partial and asymmetric modernization' and its particular brand of chaotic neoliberal capitalism, as well as endemic corruption (found to be even worse than in Greece). The authors conclude by making policy suggestions for how the Albanian government could do more to both help returnees reintegrate and capitalize on their development potential.

Notes

1. For a policy-focused overview of recent research on European migration see R. King and A. Lulle, *Research on Migration: Facing Realities and Maximising Opportunities*, Publications Office of the European Union, Luxembourg, 2016.
2. There is copious and scattered literature on individual countries, and other studies and collections which look at the Western Balkan region as part of wider geographical assemblage such as the Balkans as a whole, or South-East Europe. See for example H. Vermuelen, M. Baldwin-Edwards and R. van Boeschoten (eds), *Migration in the Southern Balkans: From Ottoman Territory to Globalized Nation States*, SpringerOpen, Cham, 2015; M. Zbinden, J. Dahinden and A. Efendic (eds), *Diversity of Migration in South-East Europe*, Peter Lang, Bern, 2016; and R. King, M. Povrzanovic Frykman and J. Vullnetari (eds), *Migration, Transnationalism and Development in South-East Europe and the Black Sea Region*, Routledge, London, 2017.
3. This was the 2nd Annual Conference of the Western Balkans Migration Network, on the theme 'Migration in the Western Balkans: What Do We Know?' The conference took place on 19–20 May 2017, hosted at International Burch University Sarajevo, and financially supported by the EU's Central European Initiative, the Konrad Adenauer Stiftung, and the government of Bosnia and Herzegovina (Ministarstvo za ljudska prava I izbjeglice).
4. See IOM, *World Migration Report 2018*, IOM, Geneva, 2018, p. 13.
5. For a thorough and balanced overview of the migration-development debate see H. de Haas, 'Migration and development: a theoretical perspective', *International Migration Review*, 44(1), 2010, pp. 227–264.

6. This has been especially and convincingly demonstrated for the case of Albania. See for example N. de Zwager, I. Gedeshi, E. Germenj and C. Nikas, *Competing for Remittances*, IOM, Tirana, 2005; and J. Vullnetari and R. King, *Remittances, Gender and Development: Albania's Society and Economy in Transition*, I.B. Tauris, London, 2011.
7. See, for example, Y. Shain, 'The role of diasporas in conflict perpetuation and resolution', *SAIS Review*, 22(2), 2002, pp. 115–144; S. Carter, 'The geopolitics of diaspora', *Area*, 37(1), 2005, pp. 54–63; R. Koslowski, 'International migration and the globalization of domestic politics: a conceptual framework', in R. Koslowski (ed.) *International Migration and the Globalization of Domestic Politics*, Routledge, New York, 2005, pp. 5–32; and M. Koinova, 'Diaspora mobilisation for conflict and post-conflict reconstruction: contextual and comparative dimensions', *Journal of Ethnic and Migration Studies*, 44(8), 2018, pp. 1251–1268.
8. R. Ramamurti, 'Developing countries and MNEs: extending and enriching the research agenda', *Journal of International Business Studies*, 35(4), 2004, pp. 277–283.
9. As a result, such return migration may result in psychosocial hardship. See S. Porobić, '"Invisible" returns of Bosnian refugees and their psychosocial wellbeing', in Z. Vathi and R. King (eds), *Return Migration and Psychosocial Wellbeing: Discourses, Policy-Making and Outcomes for Migrants and their Families*, Routledge, Abingdon, 2017, pp. 108–125.
10. R. King, 'Across the sea and over the mountains: documenting Albanian migration', *Scottish Geographical Journal*, 119(3), 2003, pp. 289–309.
11. Ibid.
12. M. Uvalic, *Science, Technology and Economic Development in South Eastern Europe*, UNESCO Science Policy Series 1, Paris, 2005.
13. In addition to the paper by Kerpaci and Kuka in this issue, see: L. Labrianidis and A. Lyberaki, 'Back and forth and in-between: Albanian return migrants from Greece and Italy', *Journal of International Migration and Integration*, 5(1), 2004, pp. 77–106; L. Labrianidis and P. Hatziprokopiou, 'The Albanian migration cycle: migrants tend to return to their country of origin after all', in R. King, N. Mai and S. Schwandner-Sievers (eds), *The New Albanian Migration*, Sussex Academic Press, Brighton 2005, pp. 93–117; L. Labriandis and B. Kazazi, 'Albanian return migrants from Greece and Italy: their impact upon spatial disparities within Albania', *European Urban and Regional Studies*, 13(1), 2006, pp. 59–74; and E. Germenji and L. Milo, 'Return and labour status at home: evidence from returnees in Albania', *Southeast European and Black Sea Studies*, 9(4), 2009, pp. 497–517.
14. Porobić, op. cit.
15. S. Bagshaw, 'Benchmarks or Deutschmarks? Determining the criteria for the repatriation of refugees to Bosnia and Herzegovina', *International Journal of Refugee Law*, 9(4), 1997, p. 592. On the wider issue of return migration in the Balkans see the collection of papers in R. Black, M. Eastmond and S. Gent (eds), 'Sustainable return in the Balkans: beyond property restitution and policy', *International Migration*, 44(3), special issue, 2006..
16. For Albania: World Bank, *Migration and Remittances Factbook*, World Bank, Washington, DC, 2011, p. 54; For Bosnia: H. Halilovich, *Mapping BiH Diaspora in 10 Destination Countries*, IOM and MHRR Report, Sarajevo, 2018.
17. Most Western Balkan countries allow their citizens abroad to vote in elections in their home country.
18. A. Paasi, 'Regions are social constructs, but who or what "constructs" them? Agency in question', *Environment and Planning A*, 42(10), 2010, pp. 2296–2301.
19. As Prelec points out, the empirical evidence for this hypothesis is not very robust, especially for the Balkans. One study which is relevant is Sean Carter's paper on the Croatian diaspora in the United States. Carter demonstrated their strong support (moral and financial rather than electoral) for the independence of Croatia during the break-up of Yugoslavia and the Balkan conflicts of the 1990s, based on essentializing narratives of Croatian-ness and nationalist self-determination. See Carter, op. cit.
20. Even if, in late August 2018, the successful completion of the financial rescue programme for Greece was finally announced, the bailout has left the country's economy and social

welfare system potentially crippled over the much longer term. For a trenchant critique of Greece's austerity politics and the wider financial architecture of the EU and IMF which shaped the rescue package, see Y. Varoufakis, 'Greece was never bailed out—it remains a debtor's prison and the EU still holds the keys', *The Observer*, 26 August 2018, p. 40.

21. R. King, 'Albania as a laboratory for the study of migration and development', *Journal of Southern Europe and the Balkans*, 7(2), 2005, pp. 133–155.

Disclosure statement

No potential conflict of interest was reported by the authors.

From Source to Corridor: Changing Geopolitical Narratives about Migration and EU-Western Balkans Relations

Ferruccio Pastore

ABSTRACT

The 'refugee crisis' of 2015–2016 brought the 'migration factor' back to the centre of the political agenda in European Union–Western Balkans (WB) relations. But the way in which population movements from and through the region are framed by the EU has changed deeply from the past. After having been seen in the 1990s as the main sending region for irregular and humanitarian migration flows to Western Europe, and following a period of decreasing political attention in the 2000s, the opening of the 'Western Balkans route' for forced migrants from the Middle East turned the region into a 'transit corridor'. Drawing upon literature on the role of policy frames and narratives in migration policy-making, the article reconstructs this evolution, identifying a trend towards narrative simplification based on an essentialized understanding of the relations between migration dynamics and territories. The risks of 'geopolitical determinism' in shaping future strategic choices are pointed out.

Introduction: the geopolitical turn in EU's migration policies

Often dismissively labelled as an economic giant but a political dwarf,[1] the European Economic Community (EEC, rebranded European Community—EC—in 1993) long prospered under the protective arm of the North Atlantic Treaty Organization (NATO), which allowed it to avoid taking direct responsibility of its own international security. Following World War II, West European political elites were reluctant to think of the region and its surroundings in explicit geopolitical terms. In part, this was due to the tainted memory of geopolitics as a discipline that had acted as a sort of 'intellectual poison' in the continent's recent totalitarian past.[2]

This started to change from the late 1980s and, at an accelerated pace, in the early 1990s, when a perception of growing instability in their common neighbourhood induced the EC's member states to establish (with the 1992 Maastricht Treaty) a European Union. To face the challenges of the new international environment, the EU was provided with new ad hoc institutional branches (called 'pillars') dealing respectively with 'Common Foreign and Security Policy' (CFSP) and with 'Justice and Home Affairs' (JHA). In that new context, the traditional geopolitical 'shyness' of European states and institutions was gradually overcome. As put by Luiza

Bialasiewicz: 'European media and politicians have become visibly less reticent to speak openly of a "European geopolitics"—or certainly of the need for a geopolitical vision for (EU)rope.'[3]

Although political concerns associated with migratory flows were from the outset a key driver of the major institutional changes agreed upon in Maastricht, migration issues were at the time essentially framed as 'internal' and therefore encapsulated in the policy domain traditionally defined as 'home affairs'. It is only at the end of the 1990s that an 'external dimension' of European migration policies emerged and migration issues started to contribute more explicitly and decisively to European geopolitical visions and strategy-making.[4] In 2005, the deadly shooting of 14 African migrants attempting to climb over the border fences between Morocco and the Spanish enclaves in Ceuta and Melilla triggered a further institutionalization and reinforcement of the external dimension of European migration policies, with the launch of the Global Approach on Migration (GAM).[5]

The (actually modest) surge in cross-Mediterranean unplanned migration, mainly from Tunisia and Libya, caused by the 2011 'Arab Spring' brought a further upgrade of the EU's foreign migration policy.[6] But it was not until 2015 that migration took a central role in shaping the European Union's geopolitical frames and priorities. The trigger event was a further, and this time more significant increase in irregular arrivals to Greece and Italy, and the outburst of what was alternatively defined as a 'refugee' (by far the most prevalent definition in the global public debate and in the media), governance or legitimation crisis.[7]

Since 2015, the overall strategic goal of a more effective migration management (a euphemism for reduction and prevention of unplanned and irregular migration flows) has become one of the key parameters driving the EU's foreign policy and external action at large. As explicitly stated by the Commission in its landmark June 2016 blueprint for the reform of cooperation with sending and transit countries, no policy area is exempt:

> Increasing coherence between migration and development policy is important to ensure that development assistance helps partner countries manage migration more effectively, and also incentivises them to effectively cooperate on readmission of irregular migrants. … The same should be true of trade policy … Even though considered potentially effective, and building on their positive impact on migration, neighbourhood, development and trade are not the only policies that are relevant to support the compacts. No policy areas should be exempted from this approach. All EU policies including education, research, climate change, energy, environment, agriculture, should in principle be part of a package, bringing maximum leverage to the discussion.[8]

As observed by Michael Collyer and Russell King, this shift of policy paradigm has been strongly inspired and driven by geopolitical frames and narratives:

> The observation that geography is central to the current crisis has become commonplace in media discussions. … More-substantial interpretations of political geography involve the central geopolitical nature of EU migration governance … the crisis is a geopolitical one rooted in more-fundamental divisions in Europe.[9]

While this is certainly true, it is crucial to add that the geopolitical approach driving this new season of EU migration policies has become more selective and more contingent than in an even recent past. Compared to the global and rather undifferentiated focus of the GAMM, and even to the European Agenda on Migration presented by the Commission in May 2015 (COM(2015) 240 final), the 'new Partnership Framework with third countries under the European Agenda on Migration' (COM(2016) 385), endorsed by the European Council in June 2016, showed a much higher level of geopolitical prioritization. As a matter of fact, the primary targets of the Partnership Framework were a handful of countries in sub-Saharan Africa (Ethiopia, Mali, Niger, Nigeria and Senegal). These few countries were singled out on the basis of their perceived role as main sources or transit corridors for 'mixed migration flows' (including both refugees and other categories of migrants) potentially heading towards Europe through Libya, along what European institutions have labelled the 'Central Mediterranean route'.[10]

But is essential to underline that it was possible for Brussels to concentrate its most recent 'migration management' efforts in sub-Saharan Africa only after the 'Western Balkans route', through which more than a million asylum seekers had entered the EU in 2015, had been virtually closed through a series of coordination measures (see below) and eventually by the controversial March 2016 'deal' with Turkey.[11]

Although nowadays somewhat overshadowed by the current focus on Africa, the Balkans remain an important element in the new, largely migration-driven, geopolitical *Weltanschauung* of the European Union. As a matter of fact, the 'refugee crisis' of 2015–2016 has brought the 'migration factor' back to the core of EU–Balkans relations. But the way in which migrations from and through the Western Balkans (WB) are framed in and by the European Union has deeply changed from the past: from a period (1990s) in which the region was seen as the main source of irregular and humanitarian migration flows to Western Europe, through a period of decreasing political attention in the 2000s, until the recent emergence of a 'transit frame' essentializing the WB as a potential corridor for forced migrants from the Middle East.

The next two sections will be devoted to a reconstruction of the evolution of the EU's geopolitical narratives about migration in, through and from the WB region over the last quarter of a century. In the subsequent section I will suggest an interpretation of these long-term developments, identifying a trend towards the essentialization of each country's 'migration functions' in the framework of an increasingly geopolitical understanding of migratory phenomena by EU institutional actors. Some normative concluding reflections will highlight the risk of such a geopolitical determinism for the future of EU–WB relations.

The EU's changing geopolitical narratives about migration from and through the WB

The early 1990s saw what some authors called an 'argumentative turn' in different areas of political studies, from policy analysis[12] to international relations.[13] A new focus on the role of discursive frames and narratives in policymaking has also been developing in the field of migration policies, most significantly since the 2011 work by Boswell, Geddes and Scholten.[14] Their particular interest in 'the very specific role played by knowledge claims in making policy',[15] is also what drove Boswell and her colleagues' choice of the concept of 'policy *narrative*' as key analytical tool (as distinct, for instance, from the widely used concept of 'policy *frame*'). Here is how the authors explained their terminological and conceptual decision:

> The focus on the factual or cognitive content of narratives distinguishes it from 'discourses', 'frames' or 'paradigms', which capture a more general perspective or set of ideas. This precision, and the emphasis on cognitive content, makes the concept of narratives better equipped to shed light on how certain factual claims can justify quite specific courses of action.[16]

I will follow along the same track, but with an important variation: my main focus will not be on the specific causal contents of migration policy narratives (*What is the problem? What causes the problem? What are the solutions?*), as in most of the articles collected by Boswell et al. My core focus will instead be on the specifically geopolitical dimension of the EU's narratives about migration in, through and from the WB, and on the variations in such narratives over the last quarter of a century. My driving empirical questions will thus be the following:

(1) Have the WB, over this period, been perceived and represented by EU institutions as a foreign policy priority *for reasons associated with international migration*?

(2) To the extent that the answer to the first question is affirmative, what has been—from a migration point of view—the perceived role of the WB region as a whole (or possibly only of some its countries) in its relations with the EU and its member states? Have WB countries been viewed and depicted (and therefore treated) mainly as 'sending', 'receiving' or 'transit' countries, or as some sort of combination of these stylised migration functions?[17]

The empirical picture presented here is exploratory, as it is based on the results of a carottage and inductive content analysis focusing on a selection of EC/EU key documents, with particular regard to those produced by the organization's apical body, i.e. the European Council.[18] This limited empirical focus certainly does not ensure a systematic coverage of the migration-related geopolitical narratives of European institutions towards the WB. But given the essentially conceptual goals of this article, narrowing the focus to key documents and particularly to the European Council's Conclusions makes specific sense as a way to intercept *only discourses and narratives that have made their way to the top of the EU's agenda.* Also the absence of frames associating the WB and migration at this level will thus be interpreted as a negative revealer of salience and therefore also considered meaningful.

From Gastarbeiter-providers to refugee-sending countries

During the Cold War, Yugoslavia never experienced the kind of demographic isolation that was common for other countries in the socialist bloc. On the contrary, particularly since the signing of a labour agreement with the Federal Republic of Germany (FRG) in October 1968, the Yugoslav Republic became an important supplier of migrant workers for the *Wirtschaftswunder*. It is estimated that, at the end of the 1980s, of the around 1.8 million *Gastarbeiter* ('guestworkers') in the FRG, more than 340,000 were from Yugoslavia.[19] But in spite of its importance, the quiet flow of guestworkers from

Yugoslav rural areas and cities to German factories remained at the level of bilateral relations, without affecting the institutions of the EEC, which at the time still lacked an international projection.[20]

It was only after the fall of the Berlin Wall that migration from the WB started to change nature in the eyes of Western Europeans. The first strong signal came from Albania, until then by far the most isolated (also from a human mobility point of view) of the European socialist regimes. In July 1990, several Western embassies in Tirana were peacefully besieged by thousands of protection-seekers, a large number of whom obtained visas, and ultimately asylum, mainly in Germany, Italy and France. Later that year, some 3000 Albanian nationals crossed the land border with Greece without meeting any resistance from Albanian border guards. But this was only a small preview of what would happen the following year. By mid-March 1991, 20,000 Albanians, many of them belonging to the Greek Albanian minority, had entered Greece, where most were allowed to remain on a temporary residence permit. In the meantime, between 7 and 10 March, another 20,000 Albanians managed to reach the Italian shores from Durrës by using all kinds of (mostly unseaworthy) vessel. Later that year, after the first multi-party elections were held and a UN interagency mission had assessed that the great majority of those who left were 'seeking better economic conditions', a new wave of arrivals by sea was met by Italian authorities with a completely different treatment. The majority of the estimated 10,000 who had disembarked on 8 August were detained in a stadium in Bari and then forcibly repatriated.

The 1991 events marked a turning point in the history of Albanian migration, with an abrupt categorical shift from a humanitarian narrative to a security-based one. Almost literally *dans l'espace d'un matin*, Albanians turned from deserving protection-seekers into threatening irregular migrants.[21] Even in 1997, when the collapse of the so-called 'Pyramid schemes', in which almost two-thirds of the population had invested, led the country to the brink of civil war, the fundamental policy frame with which the huge migration pressure from Albania was tackled remained the same. Unauthorised outflows kept being labelled and treated as illegal and only limited opportunities for managed labour and family migration to Italy were left.[22]

Given this article's focus on European narratives, what is particularly worth stressing, however, is the lack of specific attention shown at the time by European institutions. While in 1991 European Council Conclusions remained completely silent about the dramatic Albanian transition, not to speak of its sizeable migration implications, in 1997 European leaders did not go beyond welcoming 'the important role of the Multinational Protection Force in helping to create a secure environment for the election process'.[23]

Both the prevalent legal-policy frames and the overarching narrative changed again, quite dramatically, in the years following the first Albanian exodus, when the most salient migration-related events in the Balkans became the massive internal displacement and international refugee flows caused by the violent dissolution of former Yugoslavia. In that context, the humanitarian frame became absolutely dominant, in line with the wider policy of European institutions, that proved unable to give political (and, even less, military) responses to the spreading of the conflict and limited themselves to the provision of humanitarian aid in the region. In 1992, for instance, the peak year for refugee arrivals from Bosnia-Herzegovina in Western Europe, the

European Council declared itself 'gravely preoccupied by the increasingly intolerable situation of the hundreds of thousands of displaced persons in the territory of the former Yugoslavia' and entrusted the Commission to coordinate enhanced humanitarian efforts.[24]

With regard to asylum policies as such, although the war in former Yugoslavia did not immediately trigger a common European response, it was critical in *indirectly* promoting the first developments towards harmonisation of member states' national policies. In that process, a decisive role was played by Germany which, as the destination of the vast majority of war refugees (in 1992 alone, it received 438,191 applications out of a total of 674,056 for all of the EU15), was keen to promote convergence among national asylum systems as a way to achieve better 'burden sharing' (*Lastenverteilung* in German), according to the de-humanising terminology that started to permeate the policy debate at the time.[25]

The humanitarian framing was even more predominant in 1999, when the war in Kosovo generated a massive wave of forced migration (600,000 refugees and 400,000 internally displaced persons, amounting to around half of Kosovo's total population). In that dramatic context, the humanitarian narrative gained such a centrality as to become the key argument in support of the NATO-led military intervention against Belgrade. This is reflected with particular emphasis in the 'Statement concerning Kosovo' adopted by European leaders in March 1999:

> On the threshold of the 21st century, Europe cannot tolerate a humanitarian catastrophe in its midst. It cannot be permitted that, in the middle of Europe, the predominant population of Kosovo is collectively deprived of its rights and subjected to grave human rights abuses. We, the countries of the European Union, are under a moral obligation to ensure that indiscriminate behaviour and violence ... are not repeated. We have a duty to ensure the return to their homes of the hundreds of thousands of refugees and displaced persons.[26]

The international and European policy response was mainly aimed at providing humanitarian assistance and protection *in the region*. However, it is worth recalling that a vast international wave of public solidarity prompted the Western allies to launch, on 4 April 1999, a Humanitarian Evacuation Programme (HEP) through which almost 100,000 refugees were airlifted from camps in Macedonia to Germany, the United States, Turkey and 25 other (mainly European) countries.[27] Several observers have contrasted the relative generosity displayed by Western countries with that large-scale resettlement operation with the more restrictive approaches adopted on other occasions, before and (even more so) after 1999.[28]

Oscillations in salience and narrative shifts

The Kosovo crisis marked a turning point in EU–WB relations, both in general political terms and with particular regard to migration policy. Brussels' approach to the region was upgraded with political (and migratory) stabilization as a strategic goal, and with EU accession as a long-term prospect. In June 1999, the Stabilization and Association Process (SAP) was adopted as the EU's overarching policy towards the region, accompanied by the establishment of the Stability Pact for South-Eastern Europe, a regional cooperation forum (superseded in 2008 by a 'Regional Co-operation Council').

From a sectoral point of view, an important development took place in 2002, when the Stability Pact promoted a Migration, Asylum, Refugees Regional Initiative (MARRI) involving its member states, international organizations and NGOs. The establishment of this new forum paved the way for a gradual reinforcement of intra- and inter-regional cooperation in the migration and asylum policy field. The driving thematic priorities, largely dictated from Brussels, were readmission of undocumented migrants, reintegration of refugees, institution building and policy development (starting from border controls and the countering of smuggling and trafficking).

The feeling that the stabilization process was progressing—including in the migration field—was confirmed by the EU-WB Summit held in Thessaloniki on 21 June 2003. In the summit's final Declaration, EU and WB leaders reiterated their 'unequivocal support to the European perspective of the Western Balkan countries [whose] future ... is within the European Union'.[29] Once more, the EU side made its migration-related priorities explicit:

> Sustainable return of refugees and internally displaced persons is critical for ethnic reconciliation and an index of democratic maturity; it remains high on our priority agenda. ... Particular attention will be given in combating trafficking in human beings. The countries of the region also commit to concrete measures ... in order to cope effectively with illegal immigration and improving border security and management, aiming at achieving European standards.[30]

In the meantime, the two rather unequal blocs of partners acknowledged

> [the] importance the peoples of the Western Balkans attach to the perspective of liberalisation of the EU's visa regime towards them [recognising] that progress [in that area] is dependent on implementing major reforms in areas such as the strengthening of the rule of law, combating organised crime, corruption and illegal migration, and strengthening administrative capacity in border control and security of documents.[31]

The agenda for the following years was thus clearly set. In a climate of decreasing anxiety about 'migratory pressure' from the WB, and with the EU's attention increasingly focused on the unexpectedly large migratory impact of the two 'Eastern enlargements' of 2004 and 2007, in 2008 visa liberalization dialogues were launched with little public debate with Albania, Bosnia-Herzegovina, Macedonia, Montenegro and Serbia. A few months later, 'Visa Liberalization Roadmaps' were presented by the European Commission to the five WB governments. Relatively quick negotiations led to the lifting of visa obligations for the citizens of Montenegro, Serbia and Macedonia since December 2009, and of the two remaining countries one year later.

Soon after, however, the number of asylum claims laid in Western European countries by WB nationals started to increase beyond expectations. New asylum claimants from Serbia (including Kosovo) rose from 22,342 in 2009 to 40,912 in 2010.[32] A similarly steep growth, although with lower absolute numbers, was registered in 2011 for citizens of Albania (3916, up from 2571 in 2010) and Bosnia and Herzegovina (from 1758 to 3224). In the following years the trend grew exponentially: in 2015, the new applications had reached 73,987 for Albanians and 109,904 for Serbians/Kosovars (of which there were respectively 53,805 and 50,127 in Germany alone).

This surge in the number of asylum applications, most of which were eventually rejected, led to rapid policy reactions, at both EU and national level. The most important counter-measure taken at supranational level was the reinforcement of an already existing 'visa suspension mechanism'. The revised suspension mechanism (proposed by the Commission in May 2016, adopted in February 2017 and entering into force on 28 March 2017) now allows not only the member states but also the Commission itself to temporarily suspend the visa exemption for the nationals of a given third country in case of a substantial increase in irregular migration or of a specific internal security risk that may arise from visa-free travel.

As for national measures, the most consequential one was taken by the German government, soon followed by other EU executives. Between 2014 and 2015, through two subsequent bills, all WB countries were included in Germany's national list of 'safe countries of origin'.[33] As a consequence, applications by WB nationals are since then considered 'manifestly unfounded', unless the applicants provide very strong evidence demonstrating that they were personally persecuted in spite of the generally non-problematic human rights situation in their country of origin.

The official categorisation of WB countries as 'safe' marked a crucial shift in migration policy framing and allowed the construction of a consistent and powerful narrative, portraying WB countries as 'normal' European countries, whose citizens can a priori be labelled as probably *mala fide* if they ever ask for international protection.

The opening and closure of the 'Western Balkans route'

The steep growth in asylum applications by nationals of some WB countries since 2010 brought the region back onto the EU's migration policy radar. But this increase in salience was just a minor oscillation compared to what happened in the summer of 2015 when, in the space of just a few weeks, due to a completely unforeseen combination of factors (reduced exit controls by Turkey; poor border management by Greece; a 'magnet effect' generated by the German suspension of Dublin rules in August 2015), the region turned from a reactivating sending basin into the largest transit corridor for unplanned 'mixed migration' flows to the European Union.

The geopolitical metaphor that quickly imposed itself (also in the EU's official language) to describe this development has been that of a 'Western Balkans route'. As a matter of fact, the singular ('route') is misleading. The over 2 million irregular border crossings cumulatively reported by all WB countries in 2015[34] followed a variety of different trajectories, determined and constantly reshaped by a complex (and still under-researched) interaction between migrants' (and smugglers') agency and states' policy responses. Consistent with the region's pre-existing migration dynamics, however, two main itineraries emerged:

Itinerary A: Turkey → Bulgaria → Serbia → Hungary.
Itinerary B: Turkey → Greece → Macedonia → Serbia → Hungary/Croatia.

Both starting from Turkey, the two itineraries diverged in their central part, to bifurcate again at the exit from Serbia. As far as Itinerary B is concerned, this final bifurcation created two variants that I will call Itinerary B1, heading straight north from

Figure 1. The main migrant itineraries composing the 'Western Balkan route'.
Map by the author, based on Wikimedia Commons. Main source for the contents: Frontex, 'Western Balkans. Annual Risk Analysis', Warsaw, 2016.

Serbia to Hungary, and Itinerary B2, swerving to the west across Bosnia-Herzegovina and/or Croatia and then north again to Hungary or Slovenia (see Figure 1).

In the dramatic months of 2015–2016, a series of shifts among these different routes could be observed. Itinerary A was probably the most frequently used until August 2015, when Bulgarian authorities completed the construction of a fence along the border with Turkey, which they had started erecting in 2013.[35] Flows were then largely redirected along Itinerary B1, which prevailed until mid-September 2015, when Hungary sealed its border with Serbia thus pushing migrants along Itinerary B2. When the border with Croatia was also fenced off by the Hungarian authorities on 15 October 2015, the movement was redirected again, this time towards the Croatian–Slovenian border.[36]

This reckless domino game, based on a chain of 'beggar-thy-neighbour' measures, reached a level where it clearly risked reactivating the worst demons of the region, soothed but not extinguished. This dynamic raised serious anxiety at all levels, from the region's local communities to the highest policy circles in the EU and in European capitals. The alarmed words of EU Commissioner Johannes Hahn, in a speech given on 17 September 2015, are revealing about the climate of those days:

> Let's be crystal-clear: the Western Balkans are an enclave in the EU. To get there, a refugee has to walk through at least one Member State. We can't blame them for lax border controls if our own Member States don't do their job. [...] we have no interest in adding

> instability. [...] All must shoulder their fair share—and these countries are willing—but the Western Balkans must not become a parking lot or 'no man's land' for stranded refugees. That would be a grave geostrategic mistake![37]

This disruptive policy pattern started to be reversed with a so-called 'mini-summit' held on 20 October 2015 in Brussels, when eight member states (Austria, Bulgaria, Croatia, Germany, Greece, Hungary, Romania and Slovenia) and three candidates (Albania, Macedonia and Serbia) reached an agreement on a 17-point plan of action, backed by generous financial support provided by international and regional financial institutions. Coordinated implementation progressed in the following weeks, but a decisive turning point came only in March 2016, when Macedonia closed off its border with Greece, with full political and operational support from a number of EU member states (the four countries of the Visegrád Group, plus Austria and Croatia) and less vocal but decisive support coming from Brussels.

This proved effective and allowed European Council President Donald Tusk to tweet triumphantly on 9 March 2016: 'Irregular flows of migrants along Western Balkans route have come to an end. Not a question of unilateral actions but common EU28 decision.' In fact, however, the real closure of the WB route came about only a few weeks later, when new arrivals in the EU were prevented through the costly and controversial deal struck with Erdoğan's Turkey.

To sum up the empirical developments described in this and the previous section, the way in which EU institutions have perceived migration from and through the WB has changed dramatically over the last quarter of a century, in terms both of levels of salience and of overarching geopolitical narratives (with the region being constructed mainly as a sending basin or a transit area). These diachronic changes are schematically represented in Figure 2.

Essentialization of migration functions and the risks of geopolitical determinism

In sharp contrast with the well-known optimistic narratives on post-Cold War globalization as not just a post-historical but also a post-geographical era, it soon became clear that, even though in 1989 an important wall had fallen, borders were not dead. On the contrary, while their functions unquestionably changed,[38] their symbolic and practical importance even grew, pushing further an already existing tendency towards their 'overdetermination'.[39] With particular regard to the European context, Jan Zielonka describes a process of 'unbounding and rebounding' at the end of which, however, the political importance of borders is firmly re-established.[40]

A growing securitized perception of international migration has undoubtedly been a critical driver of this overdetermination of borders and, more generally, of an ever more essentialized understanding of the relation between territory and human mobility. Geopolitical frames and narratives have given a critical contribution in this direction. As Jennifer Hyndman puts it: 'Geopolitical scholars have essentialized "the territorial" at the expense of examining how power relations render specific bodies objects of surveillance and discipline.'[41]

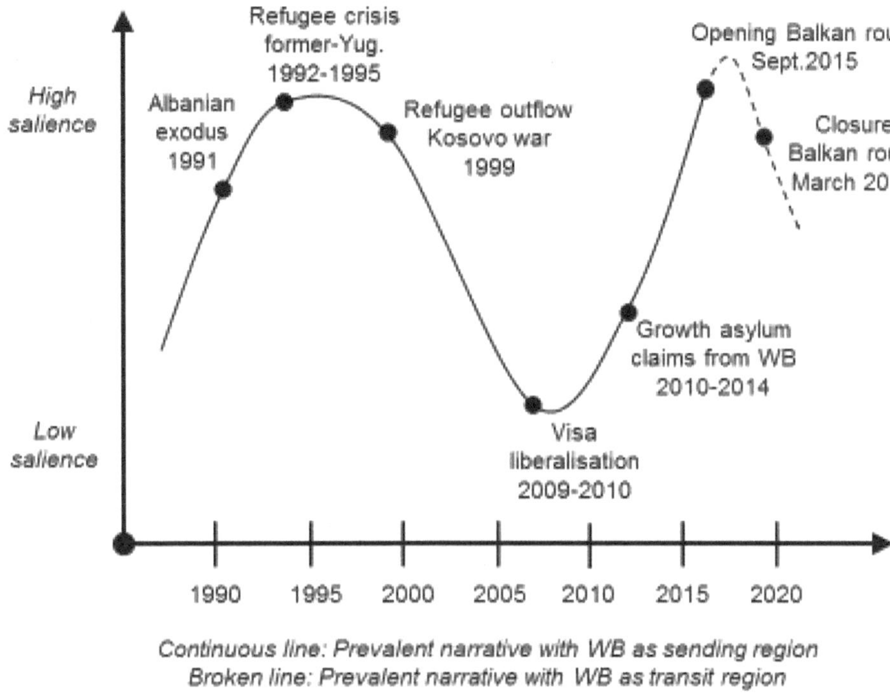

Figure 2. Main events shaping EU policy narratives (and associated priority levels) about migration from/through the Western Balkans (1990–2018).

Besides the transformations in the political and symbolic functions of international borders, processes of overdetermination and essentialization are also witnessed at other scales. In particular, a tendency can be observed to rigidly categorize countries (as both geographical and political entities) into one of the three basic 'migration functions' that a given territory can in theory play: as 'sending', 'receiving' or 'transit' spaces.[42]

As anticipated above, this categorization is usually based on highly stylized descriptions of actual migration dynamics, which tend to disregard their intrinsic complexity and lack of linearity, both in spatial terms (with 'stepwise' or 'fragmented' migration trajectories becoming more common, or at least increasingly thematized by researchers[43]) and in temporal terms (with non-linear patterns such as temporary returns, circular migration, etc., becoming more frequent or, at least, increasingly thematized). In this kind of simplified geopolitical migration narratives, the idea that a single local or national space can simultaneously perform different migration functions, or that its main migration function (as a sending, receiving or transit space) can rapidly and repeatedly change over time, is often neglected.[44] Territories are thus reduced to their prevalent migration function or, better, to the migration function that is perceived as most relevant (in positive or, more frequently, in negative terms) for the political or economic interests of the subject who is producing (and making use of) the narrative (in this case, European institutions).[45] Furthermore, the only territorial unit of analysis considered viable and pertinent is usually the nation-state, with the result that a country as a whole will be 'sending' (or 'receiving' or 'transit'), thus disregarding internal differentiations, however deep and structural.[46]

As demonstrated by the empirical developments reconstructed in the core sections of this paper, processes of essentialistic narrative simplification have routinely been used by European institutions to apprehend, represent and communicate the migration realities of the WB and the related policy challenges, with a constant and very close interplay—sometimes, one might say, a deliberate confusion—between the cognitive and the normative dimension.

Besides being a mode of migration governance that deserves more scholarly attention, this geopolitically driven migratory essentialism has important practical implications, that seem particularly evident in the case of the WB. Pinning down an entire country or region to its alleged nature of 'sending space' (through such blurred—but routinely used—policy concepts such as 'migratory risk', 'pressure' or 'potential') can generate a geopolitical trap by condemning that particular country or region to a marginal position in the global geography of human mobility, with clear prejudice to its development prospects. But, for a given territorial political entity, also being rigidly and exclusively labelled (and treated) as a 'transit space', besides some potential short-term benefit, can also have seriously detrimental long-term effects.[47] In the case of the Balkans, for instance, having performed for centuries the role of a geopolitical buffer between Christianity and the Ottoman empire, has certainly not contributed to the political stability and the autonomous and sustainable development of the region. This historical past should provide enough lessons about the risks of transforming the WB into a sort of 'migration *krajina*' (a word meaning 'border region' in several Slav languages).[48]

In its recently announced strategy for the region, the European Commission seems aware of such risks. Conscious of the 'interlinkage and interdependence of the region with the European Union [*that were*] evident in the migration crisis',[49] the EU's executive body designs an inclusive strategy, where the reinforcement of the region's migration management capacity is indissociable from regional cooperation, migrant and refugee rights, and a clear path towards EU accession. It remains to be seen if this bold and confident blueprint will gather sufficient consensus from governments, in the EU and in the region, to be consistently and energetically implemented over the coming years.

Notes

1. The attribution of the phrase is controversial. Amongst the first to use it was, in 1991, Mark Eyskens, then Foreign Minister of Belgium (as quoted by Craig R. Whitney, 'War in the Gulf: Europe; Gulf fighting shatters Europeans' fragile unity', *The New York Times*, 25 January 1991).
2. K. Dodds, *Geopolitics. A Very Short Introduction*, Oxford University Press, Oxford, 2007, p. 22.
3. L. Bialasiewicz, 'The new political geographies of the European "neighborhood"', *Political Geography*, 28(2), 2009, p. 79.
4. C. Boswell, 'The "external dimension" of EU immigration and asylum policy', *International Affairs*, 79(3), 2003, pp. 619–638; S. Lavenex, 'Shifting up and out: the foreign policy of European immigration control', *West European Politics*, 29(2), 2006, pp. 329–350.
5. The Global Approach to Migration (GAM) was proposed by the European Commission in a Communication of November 2005 (COM(2005) 621 final) and adopted by the European Council in December that year.

6. Following an overall evaluation of the first phase of its implementation, the European Commission proposed an expansion of the GAM to include also short-term mobility issues (COM(2011) 743 final). In May 2012, the policy was thus strengthened and rebranded as GAMM (with the addition of a second 'M' for 'mobility').
7. F. Pastore, *Beyond the Migration and Asylum Crisis. Options and Lessons for Europe*, E-Book Series Aspen Italia Views, Treccani, 2017 <https://www.aspeninstitute.it/aspenia-online/article/ebook-beyond-migration-and-asylum-crisis-options-and-lessons-europe-0> accessed on 6 august 2018; see also M. Collyer and R. King, 'Narrating Europe's migration and refugee "crisis"', *Human Geography*, 9(2), 2016, pp. 1–12. Available online from Sussex Research Online: <http://sro.sussex.ac.uk/68634/> accessed on 6 august 2018. The EU's cognitive and normative frames about the Mediterranean represent the core focus of an ongoing collaborative research project called Medreset (http://www.medreset.eu). On European policy frames about migration, see also J. Carling and M. Hernández-Carretero, 'Protecting Europe and protecting migrants? Strategies for managing unauthorized migration from Africa', *British Journal of Politics and International Relations*, 13(1), 2011, pp. 42–58; M. Ceccorulli and S. Lucarelli, 'Migration and the EU global strategy: narratives and dilemmas', *The International Spectator*, 52, 2017, pp. 83–102.
8. European Commission, Communication on 'Establishing a new Partnership Framework with third countries under the European Agenda on Migration' (COM(2016) 385 final), June 2016, p. 9, <https://ec.europa.eu/home-affairs/sites/homeaffairs/files/what-we-do/policies/european-agenda-migration/proposal-implementation-package/docs/20160607/communication_external_aspects_eam_towards_new_migration_ompact_en.pdf> accessed on 6 august 2018.
9. Collyer and King, op. cit., p. 6.
10. The EU's border control agency Frontex identifies eight main 'routes', i.e. geographical trajectories typically used by migrants and smugglers for irregular access to the territory of the EU: (a) Western African route; (b) Western Mediterranean route; (c) Central Mediterranean route; (d) Apulia and Calabria route; (e) Circular route from Albania to Greece; (f) Western Balkans route; (g) Eastern Mediterranean route; (h) Eastern Borders route <http://frontex.europa.eu/trends-and-routes/migratory-routes-map/> accessed on 6 august 2018.
11. For a critical legal analysis, see S. Peers, 'The final EU/Turkey refugee deal: a legal assessment', *EU Law Analysis Blog*, 18 March 2016 <http://eulawanalysis.blogspot.it/2016/03/the-final-euturkey-refugee-deal-legal.html> accessed on 6 august 2018.
12. F. Fischer and J. Forester (eds), *The Argumentative Turn in Policy Analysis and Planning*, Duke University Press, London, 1993.
13. J. Goldstein and R.O. Keohane, 'Ideas and foreign policy: an analytical framework', in J. Goldstein and R. O. Keohane (eds), *Ideas and Foreign Policy: Beliefs, Institutions, and Political Change*, Cornell University Press, Ithaca, NY, 1993, pp. 3–30.
14. C. Boswell, A. Geddes and P. Scholten, 'The role of narratives in migration policy-making: a research framework', *British Journal of Politics and International Relations*, 13(1), 2011, pp. 1–11.
15. V Ibid., p. 2.
16. Ibid., p. 4.
17. It has become a standard approach (and indeed almost routine) in the international migration policy domain to describe countries in terms of their prevalent 'function' in international migration dynamics, the three basic functions being 'sending' (other frequently used terms are 'source' or 'origin'), 'receiving' (or 'destination'), and 'transit' country (the 'corridor' metaphor being also frequently used as an equivalent). As I will argue in greater detail in the final section, this categorization is usually based on highly stylized descriptions of actual migration dynamics, that tend to disregard their intrinsic complexity and lack of linearity.
18. European Council conclusions from 1975 onwards are available here: <http://www.consilium.europa.eu/en/european-council/conclusions/> accessed on 6 august 2018.

19. S. Mihajlovic, 'Yugoslav Gastarbeiter: the guest who stayed for dinner', *Northwestern Journal of International Law & Business*, 8(1), 1987, pp. 181–196.
20. Kaia Shonick, 'Politics, culture, and economics: reassessing the West German guest worker agreement with Yugoslavia', *Journal of Contemporary History*, 44(4), 2009, pp. 719–736, interprets West Germany's 1968 labour agreement with Yugoslavia not just as one of many similar labour market tools, but also as one of the very first moves of what in the following years would evolve into West Germany's full-fledged *Ostpolitik*.
21. For a reconstruction of Albanian migration in its different stages, see R. King, 'Across the sea and over the mountains: documenting Albanian migration', *Scottish Geographical Journal*, 119(3), 2003, pp. 283–309. For a diachronic overview of maritime irregular migration to Italy with a particular focus on cross-Adriatic flows during the 1990s, see P. Monzini, F. Pastore and G. Sciortino, 'Schengen's soft underbelly? Irregular migration and human smuggling across land and sea borders to Italy', *International Migration*, 44(4), 2006, pp. 95–119.
22. For an overview of policy responses in that phase, P. Martin, S. Martin and F. Pastore, 'Best practice options: Albania', *International Migration*, 4(1), 2002, pp. 103–118.
23. Amsterdam European Council, 16–17 June 1997, Presidency Conclusions.
24. Lisbon European Council, 26–27 June 1992, Presidency Conclusions.
25. For more details, see E. Thielemann, 'Symbolic politics or effective burden-sharing? Redistribution, side-payments and the European Refugee Fund', *Journal of Common Market Studies* 43(4), 2005, pp. 807–824. For a more general reconstruction: J. Van Selm, *Refugee Protection in Europe: Lessons of the Yugoslav Crisis*, Brill-Nijhoff, The Hague, 2007.
26. Berlin European Council, 24–25 March 1999, Statement by the European Council concerning Kosovo <http://www.consilium.europa.eu/media/21081/berlin-european-council-presidency-conclusions.pdf> accessed on 6 august 2018.
27. For an overview, see J. Van Selm (ed.), *Kosovo's Refugees in the European Union*, Bloomsbury Academic, London, 2000.
28. See, in particular, M.J. Gibney, 'Kosovo and beyond: popular and unpopular refugees', *Forced Migration Review*, 5, Special issue 'Learning from Kosovo', August 1999 <http://www.fmreview.org/sites/fmr/files/FMRdownloads/en/kosovo.pdf> accessed on 6 august 2018.
29. Eu-Western Balkans Summit, Thessaloniki, 21 June 2003, Declaration, 10229/03 (Presse 163) <http://europa.eu/rapid/press-release_PRES-03-163_en.htm> accessed on 6 august 2018.
30. Ibid., Point 6.
31. Ibid., Point 7.
32. All the figures on asylum applications are taken from UNHCR Statistical Yearbooks for 2009 and following years, Statistical Annexes <http://www.unhcr.org/search?comid=56b07b944&&scid=49aea93a5b&tid=49ec6f1727> accessed on 6 august 2018.
33. Almost simultaneously, in September 2015, the European Commission proposed (COM (2015) 452 final) to establish an 'EU common list of safe countries of origin' where all five WB countries would be included. The negotiations on this legislative proposal are currently suspended.
34. This figure, roughly 30 times more than in 2014, is reported by Frontex ('Western Balkans —Annual Risk Analysis' Warsaw, 2016, 2016, <http://frontex.europa.eu/assets/Publications/Risk_Analysis/WB_ARA_2016.pdf> accessed on 6 august 2018). The 2 million figure obviously does not correspond to the overall number of transited migrants, as each migrant on their way across the region directed north/west had to cross at least three international borders.
35. C. Mortimer, 'Bulgaria builds final part of razor wire fence to keep out refugees', *The Independent*, 4 August 2015 <http://www.independent.co.uk/news/world/europe/bulgaria-builds-final-part-of-razor-wire-fence-to-keep-out-refugees-10437962.html> accessed on 6 august 2018.

36. The reconstruction provided in the text simplifies an extremely complex dynamic that still awaits in-depth investigation. A useful chronology of migration developments on the ground is provided by S. Šelo Šabić and S. Borić, 'At the gate of Europe. A report on refugees on the Western Balkan route', Friedrich Ebert Stiftung, Zagreb, without date (but 2016) <http://www.irmo.hr/wp-content/uploads/2016/05/At-the-Gate-of-Europe_WEB.pdf> accessed on 6 august 2018; See also: 'Timeline of the European migrant crisis', in Wikipedia, the free encyclopedia <https://en.wikipedia.org/wiki/Timeline_of_the_European_migrant_crisis>. A detailed chronology of policy developments, mainly focused on the EU level, is available on the website of the EU Council <http://www.consilium.europa.eu/en/policies/migratory-pressures/history-migratory-pressures/> accessed on 6 august 2018. In 2018, apprehensions of transit migrants and asylum seekers at the border between Bosnia and Croatia started to rise again, suggesting the emergence of a new itinerary entering Bosnia from Albania and aiming at reaching the Schengen area in Slovenia (F. Pastore, 'Migrazioni nei Balcani: nuovi muri, nuove rotte', *Neodemos*, June 2018 <http://www.neodemos.info/articoli/migrazioni-nei-balcani-nuovi-muri/>) accessed on 6 august 2018.
37. Full text available at <http://europa.eu/rapid/press-release_SPEECH-15-5670_en.htm> accessed on 6 august 2018.
38. A. Kraler, M. Hendow and F. Pastore (guest editors), 'Multiplication and multiplicity: transformations of border control', *Journal of Borderland Studies*, 31(2), 2016.
39. E. Balibar, 'What is a border?', in E. Balibar (ed) *Politics and the Other Scene*, Verso, London, 2002, p. 79.
40. J. Zielonka, 'The remaking of the EU's borders and the images of European architecture', *Journal of European Integration*, 39(5), 2017, p. 644.
41. J. Hyndman, 'The geopolitics of migration and mobility', *Geopolitics*, 17(2), 2012, p. 244; see also J. Hyndman, 'Conflict, citizenship, and human security: geographies of protection', in D. Cohen and E. Gilbert (eds), *War, Citizenship, Territory*, Routledge, Routledge, 2008, pp. 241–257.
42. The essentialization of the 'transit' function and its instrumentalization for political purposes has attracted particularly intense scholarly attention. See, especially, M. Collyer, F. Düvell and H. de Haas, 'Critical approaches to transit migration', *Population, Space and Place*, 18(4), 2012, pp. 407–414.
43. Of particular importance is the work of Joris Shapendonk: see for instance J. Schapendonk, 'Turbulent trajectories: African migrants on their way to the European Union', *Societies*, 2(2), 2012, pp. 27–41.
44. The use of highly stylized migration narratives for legitimizing purposes is far from limited to the geopolitical dimension. For instance, on the functions of simplifying narratives in the field of migration control policies, see C. Boswell, 'Migration control and narratives of steering', *British Journal of Politics and International Relations*, 13(1), 2011, pp. 12–25.
45. This conventional categorization of migratory spaces mirrors the equally conventional categorization of internationally mobile persons (that in the case of the EU presents itself as a tripartition between economic migrants, refugees and other persons in need of protection, and free movers). In its positive legal dimension, the conventional categorization of internationally mobile persons is an indispensable (although contested) cornerstone of the contemporary international and European migration governance. See F. Pastore, 'The forced, the voluntary and the free: migrants' categorisation and the tormented evolution of the European migration and asylum regime', *Studi Emigrazione*, 52(200), 2015, pp. 569–586.
46. This could be seen as yet another example of the 'methodological nationalism' that Andreas Wimmer and Nina Glick-Schiller have famously denounced as a sort of genetic disorder of migration studies (A. Wimmer and N. Glick Schiller, 'Methodological nationalism and beyond: nation-state building, migration and the social sciences', *Global Networks*, 2(4), 2002, pp. 301–334).

47. For a preliminary inquiry on the potentially perverse effects of the systematic outsourcing of migration controls by European institutions to African 'sending' and 'transit' states, see F. Pastore, 'Migration and negative extraversion. Recent developments in Euro-African cooperation on migration: theoretical implications and potential effects', FIERI Working Papers, October 2017 <https://www.fieri.it/wp-content/uploads/2017/10/WP-FIERI_Migration-Negative-ExtraversionF.Pastore-UNU-WIDER-Conference-October-2017-2.pdf> accessed on 6 august 2018.
48. For a particularly pessimistic view on this point, see A. Greider, 'Outsourcing migration management: the role of the Western Balkans in the European refugee crisis', Migration Information Source, Migration Policy Institute, Washington DC, August 2017 <http://www.migrationpolicy.org/article/outsourcing-migration-management-western-balkans-europes-refugee-crisis> accessed on 6 august 2018.
49. European Commission, 'Communication from the Commission to the European Parliament, the Council, the European Economic and Social Committee and the Committee of the Regions—A credible enlargement perspective for and enhanced EU engagement with the Western Balkans', COM(2018) 65 final, Strasbourg, 6 February 2018, p. 1 <https://ec.europa.eu/commission/sites/beta-political/files/communication-credible-enlargement-perspective-western-balkans_en.pdf> accessed on 6 august 2018.

Acknowledgements

The author would like to thank Raffaella Coletti (University of Rome 'La Sapienza') for her valuable comments on an earlier version of this article. This article benefits from research carried out by the author and his FIERI colleagues Emanuela Roman and Irene Ponzo in the framework of the Medreset project (http://www.medreset.eu/), coordinated by the Istituto Affari Internazionali (IAI) and funded by the European Union's Horizon 2020 Programme for Research and Innovation under grant agreement no 693055.

Disclosure statement

No potential conflict of interest was reported by the author.

Funding

This work was supported by the European Union's Horizon 2020 Programme [Grant no: 693055].

Balkans and Baltics: On Migration as a Factor of Regional Peripheralization

Aija Lulle

ABSTRACT

This paper proposes a critical comparative reflection on migration as a factor of peripheralization processes in the Western Balkan and Baltic regions. In their respective geographical positions, the Western Balkans and the Baltics stand at different ends of the European map: south-east and north-east. Regarding their historical and political positions, however, these are very diverse: the Western Balkans were part of Yugoslavia (except Albania) while the Baltic republics (Estonia, Latvia, Lithuania) were annexed to the Soviet Union after the Second World War. Currently, Croatia is already a member of the European Union, as the three Baltic countries are, while the rest of the Western Balkans are not. The article argues that apart from economic and geographical factors, usually taken into account in describing peripheries and cores, migration processes and discourses powerfully construct some places as dynamically moving towards a 'core' while others remain confined to a 'periphery'. By paying special attention to 'scientific discourses', the aim is to broaden our understanding of theories and practices of peripheralization, where migration tends to be under-theorized.

Introduction

If we look at a map of Europe, the Baltic countries and the Western Balkans stand at two extremes, north-east and south-east, with huge distances and many different countries in between. These regions are rarely compared, and few case studies exist so far, despite the fact that there could be a multitude of historical, economic and political factors at stake to compare. In this paper I will focus on issues of migration, and more specifically on the discourses that surround and permeate the notion of migration in these regions. Such comparisons on migration discourses or practices in these regions are thus far very rare.[1] But before I approach these necessarily entangled discursive constructs, I want to briefly state my understanding of the notion of 'regions' in this paper.

One of the most influential thinkers on regions and peripheralization, Anssi Paasi, argues that regions are socially constructed, and as such should be understood in socio-spatial terms.[2] Moreover, regions are 'products that have been constructed in relation to specific criteria'.[3] These representations are always related to broad global relations

---

This article has been republished with minor changes. These changes do not impact the academic content of the article.

within which we often talk about development and migration. Through such a socio-spatial lens, the Baltics and the Balkans are not simply clusters of small states at the outer borders of the European Union, but interwoven in the globalization of neoliberal capitalism, world urbanization and other global trends. Although these geopolitical factors are pertinent and contextual, they do not fix some places into peripheries and nothing more. And, although political power relations, along with cultural representations such as 'Balkanization' or 'Eastern Europe', form discursive fields about rather weak and peripheral regions,[4] I want to leave these issues in the contextual background and concentrate on the role played by migration in peripheralization. My main focus is to bring the modes of representation of migration into the fore of my analysis of peripheralization processes.

Therefore the underpinning question of this paper is how these modes of representation of migration play a role in the construction of peripheralization. Relatedly, how do these modes construct regions (and countries within regions) as more 'peripheral' or more 'core' oriented?

From the international migration point of view, prosperous core states are more associated with the process of immigration: the economy is developing, there is a need for larger workforces, and migrants tend to move to places where they have better economic opportunities. Emigrant countries, one could assume, represent the opposite situation, associated with the economic and geographic periphery. On a national scale, cores and peripheries can be well seen through the process of urban drift: in both our case study regions there have been clear urbanization processes. Moreover, in both the Western Balkans and Baltics internal and international migration are complexly interlinked: populations have been moving towards the capital city or other urban centres, and in some cases emigration was the next step after internal migration pathways, whereas in other cases international migration prefigures internal migration when returnees originating in peripheral rural regions resettle in urban destinations.[5]

Accordingly, in the analytical sections I will focus on the above-mentioned modes of representation—emigration, depopulation, regional migration and immigration—and analyse migration in a dynamic way as shifting towards cores or peripheries as well as reshaping the definition of cores and peripheries in the process. First, however, I briefly present my theoretical framework and a note on methodology.

Theorizing peripheralization and migration

I take Paasi's arguments on regions and peripheralization as a starting point. He encourages researchers on regionalization and peripheralization to ask a question: 'Who and what constructs regions, through what associations/networks and for what purposes?', and then reminds us that 'answers are always contextual'.[6] There are many factors which have been taken into account by the various scholars who define and analyse peripheralization processes. Global factors are usually grounded in economic geography: uneven development within and between regions and states, which are lagging behind the performance as measured from the centre, the core. These economic factors are usually multiple and complex, involving macroeconomic performance, transport connectivity, accumulated capital and so on.

Regions are often simply defined and described as statistical or administrative, but they need to be critically understood as 'social spatialization': namely, as socially constructed spaces embedded in power relations, which can in turn be expressed in core–periphery frameworks. Peripheralization in the Western Balkans and the Baltics often goes hand in hand with polarization. The latter, in its basic meaning, is understood as growing differences and inequalities—economically, socially, spatially—both between countries and, more particularly, also internally. Despite the small size of all the Baltic and Western Balkan states, one of the most remarkable features of recent decades has been their ongoing socio-economic spatial polarization, and rural–urban and inter-regional migration have been intrinsic to this process. Both regions underwent profound transformations from command to market economies, privatization, growth in metropolitan and peri-urban areas;[7] and both, albeit at different times, sought access to the European Union. The Baltic states, as well as Slovenia, joined in 2004, Croatia in 2013, while the other Western Balkan states stand in a somewhat uncertain position in 2018. Hence peripheralization and polarization characterize the broader geographies of post-socialist transformation in these two groups of countries.[8] In addition, in the EU's multi-level governance framework and especially in the Eurozone, the core countries are seen as more competitive internationally as they have higher levels of productivity. By contrast, reflecting their subjection to fiscal discipline measures, peripheries have high external and public debts, as in the case of six Western Balkan countries, which Bartlett and Prica call the 'super-periphery'.[9] However, beyond economic parameters, the phenomenon of peripheralization also needs to be linked to historical and ongoing processes of migration.

For a long time in migration studies, in fact since the very beginning, *emigration* has been associated with the status of a periphery, while *immigration* has generally correlated with core regions of economic activities and better wages. *Return* migration, for its part, does not have such clear connotations. Return can be interpreted as failure to achieve pre-set goals in an immigration country, or as linked to certain life stages, for instance, after retirement. On the other hand, return can also be seen as led by innovation, and the desire to invest new knowledge and skills in the country of origin.[10] Hence, to some extent return does imply peripheralization: either as the place where 'failed migrants' return to, or as a peripheral region/country where returning migrants can potentially carry some developmental stimulus. *Transit* migration, however, is a relatively new term which can also be directly linked to peripheralization processes, especially if we see transit migration (and the ways in which regions and countries deal and cannot deal with it) through a migration governance perspective, dictated by the core.

One of the clearest recent descriptions of periphery–core relations belongs to King.[11] Through a concise review of the development of the core–periphery framework, beginning with dependency theorists[12] and world systems theory,[13] he then focuses on a lesser-known but highly relevant study of underdevelopment in Europe produced by Seers *et al.* in the 1970s.[14] Drawing on Selwyn's broad reflections on periphery,[15] King distinguishes the following characteristics. Peripheries lack effective control over many kinds of resources and innovations, whilst internal linkages within a periphery are weak. Moreover, information flows are usually from core to periphery, and likewise humans: they migrate from periphery to core regions and economies. Although these observations belong to the 1970s, they actually serve very well when reflecting on the social construction of peripheralization at the beginning of the twenty-first century, when other aspects of broad Europeanization processes are quite different. So, peripheralization can be succinctly defined as 'a loss of capacities of actors and institutions to act',[16] while hierarchies of cores and peripheries relegate the latter, as Merje Kuus has aptly summarized, to 'places of lower rank'.[17]

Both the Baltic and Balkan regions, to a certain extent, also suffer another side effect of emigration and post-socialist transformations—that of depopulation and declining birth rates. These may also result in the phenomenon of 'brain drain', reflecting limited possibilities for highly educated young people, including many recent graduates, to finding rewarding work in their home countries, encouraging them to leave in search of better income, employment and career opportunities.

In summing up the theoretical approach for this paper, I will forefront modes of representation of migration against the background of a discursive construction of peripheries as 'lagging behind', 'emptying' and 'brain draining' in these two case study regions. Simultaneously I will stay alert to alternative representations of processes of immigration and regional migration, and potentialities for development as positive counter-constructions to peripheralization. One should also keep in mind that representation is constructed by certain schools of thoughts which may, but most often do not, stem from the regions of interest of this paper. From the outside, scholars from the core regions may routinely describe these regions as peripheral, 'east', with specific contextual meanings of Balkanization[18] or post-Soviet influences.[19] Yet different narratives emerge as well. For instance, most recently Estonian researcher Maarja Saar challenged the overwhelming portrayal of Baltic emigration as labour migration and showed that lifestyle concerns can be prime motives for emigrants and returnees too.[20] This invites us to be careful and critical of how migration, as a peripheralization factor, is evaluated differently on various spatial scales, and from different viewpoints.

Note on methodology

Paasi notes that peripheralization is constructed through material and non-material elements, media representations and political communication, ranking and recommendations, as well as everyday experiences.[21] However, he emphasizes that it also blends in scientific analysis, and this 'scientific blend' is the main focus of this paper. Scientific writing, like the media but for different audiences and using different methods, can deploy the centre–periphery axis as an analytical frame at different scales.[22] Such an approach is a strategic mechanism in the construction of meaning of a periphery associated or implied with migration processes. Simultaneously, such periphery-ingrained representations further shape the geographical imagination.

Scientific representation can be made in various ways: if not using explicit theories of peripheralization, representations may draw on economic development theories, migration theories imbued with development ideas, notions of regions 'lagging behind', and of places and countries lacking capacities to act and innovate due to the draining effects of migration processes, thus emphasizing their peripherality. These narratives and discourses evolve and reshape themselves over time.

In order to trace modes of representation and their dynamics, I firstly I skimmed through a large corpus of scientific publications during the past fifteen years to get an overview of themes used when representing migration processes in the two regions. Secondly, I created a sample of 54 research papers and ordered them thematically according to the most prominent issues in both regions: (1) emigration within an overarching 'East-West' framework; (2) depopulation and brain drain; (3) specifically for the Western Balkans, transit migration and irregular border crossing; (4) weak development, notably for return migration and immigration.

Naturally, some of these themes and representations are intermingled. I also cross-referenced the relative importance of these themes with other indicators such as statistical data and media references. These four themes constitute most of the rest of this paper, following a brief historical-contextual overview.

Brief historical overview of migration processes

To understand the migration flows in the Balkans and Baltics, we need to briefly review the history of migration in these regions.[23] Before the First World War the most desirable destination of the migrants from the Balkans was the USA; whilst from the Baltics, the USA, Latin America and Russia were preferred. Two different processes took place: forced migration and migration for economic reasons.

During the interwar period the European core economies, especially France and Germany, were attractive to both Baltic and Balkan migrants, often including the middle class, students and those in creative professions. After the Second World War, large volumes of forced migration in the form of deportations from the Baltics took place, along with refugee streams towards Western Europe, some of the latter then redistributed to other Western countries globally. At least 200,000 from the three Baltic states were deported, mainly in two waves in 1941 and 1949; and even more escaped as refugees. 'Free' emigration was severely restricted to individuals, with the exception of Soviet Jews, until the end of the Soviet Union in the late 1980s.

After the Second World War, socialist Yugoslavia did not open its borders until 1963–1964, after which West Germany became the main target for emigrants. Between 1963 and 1981, around 4% of the total population left the country.[24] Return migration also took place: from 1981, about 280,000 returned but up to an estimated one-third of the returnees went abroad again, no longer able to adapt to the socialist system. Albania, on the other hand, did not allow emigration; as a consequence, its population grew fast: from 1.2 million in 1950 to 3.2 million in 1989.[25]

While in some parts of former Yugoslavia, especially in Slovenia and also Croatia, migration was less restricted and people could also move to 'Western' countries, cross-border mobility was anyway controlled with other means, including currency exchange rates. However, emigration was almost completely impossible from Albania; in fact, it was effectively the most isolated country in Europe throughout the second part of the twentieth century, until around 1990.[26] Emigration was also severely restricted from the Baltic states during Soviet times, and even travelling to other socialist countries was rather limited and a privilege, often as a reward for good work in building communist futures. Yugoslavia and the Western Balkans were seen as more developed, with a better climate and amenities that appealed to northerners from the Baltic region along with access to better goods produced outside the Soviet Union. In the meantime, there was considerable regional migration within former Yugoslavia, the Baltic states and the wider USSR, including exchanges of students and workers in the creative professions. However, nowadays the realities of mixed emigration from the Western Balkans—a recent history of forced migration due to wars, and continuing economic reasons to emigrate—pose the biggest challenges for the region's development, while population decline is an issue overshadowing the future development of the Baltics. As a result, emigration has profoundly changed these countries and regions during past two decades. Besides, the process of migration itself has changed over time, 'maturing' from temporary to more permanent forms in the Western Balkans, and especially in Albania where the scale and suddenness of mass emigration have been greatest.[27]

East–West migration

I begin with the Baltics. It is evident that local migration scholars were mainly preoccupied with the central statistical representation of the volume and structure of migration flows from the Baltic states with respect to EU enlargement.[28] Emigration increased rapidly after the Baltic states joined the EU in 2004. It was further boosted during the financial crisis in 2009–2011, especially from Lithuania and Latvia. Censuses in these countries showed that at least 213,000 left Latvia between 2000 and 2011, while 615,000 left Lithuania between 1990 and 2011.[29] The direction was towards the West. Furthermore emigration from the Baltics was represented as involving a substantial number of young and low-skilled workers.[30] 'Old' member states in the EU received large numbers of migrants from the Baltic countries, with the UK being the most important destination for Latvia and Lithuania, while most Estonians went to neighbouring Finland. Mainly young, not likely to be welfare recipients, yet working for wages which were relatively low compared to those of the local workforce, these new European migrants were a gain for EU core economies. Hence EU enlargement clearly contributed to the growth potential and output of the 'old' member states, while the human capital consequences in the Baltics remain yet to be explored in depth.[31] Moreover, even during the global economic and financial crisis, which hit the Baltic states, especially Latvia, very hard, new waves of migrants continued to make a solid economic contribution to their various destinations.[32] Namely, they were not a burden to the recipient country, while they were a human loss to rapidly depopulating regions of origin.

Albania, and Kosovo, for very different reasons and with contrasting migration dynamics, are the most extreme cases of emigration from the Western Balkans after the collapse of the communist regimes. The Kosovo war unleashed large-scale refugee movements in the 1980s–1990s.[33] This resulted in the situation where every third household in Kosovo has migrant family members, with consequent dependence on remittances and a household economy and family and social relations strongly linked to transnational ties abroad.[34] The Albanian case vies with Moldova for being the European country where the relative impact of post-communist mass migration has been greatest: according to World Bank figures, there is a 'stock' of Albanians living abroad of about 1.4 million, equivalent to nearly half of the Albanian resident population of just less than 3 million.[35] But also Serbia, Bosnia-Herzegovina and the newly named Republic of Northern Macedonia continue to show high emigration levels, albeit with strong signs of decline of emigration at the beginning of the twenty-first century.

In the 2000s representations of the Balkan model of migration have tended to portray it as occurring in stages. For instance, the fact that quantitatively emigration flows from Serbia tend to be predominantly male is interpreted as indicating that migration is still in its early stages.[36] Later studies also analysed how emigration from the Balkans normalized and matured in its trends, representing the positive dynamics of broader stabilization processes in the region. Accordingly, a stage type of representation of migration reveals novel aspects, especially with respect to how migration can contribute to processes of peripheralization. Similarly, forecasts of future emigration trends as continuous contribute to such 'peripheralist' representations, whether emigration is caused by push factors of an economic, political or nationalistic nature, or other factors.[37] Migrants from post-socialist countries are also represented as 'learning migrants', engaged in capitalist learning processes, both those searching for better economic opportunities and those facing the harsher reality of unemployment.[38]

According to the World Bank, labour markets in the Western Balkan countries are characterized by low employment rates and high unemployment by European standards. Emigration, depopulation and deindustrialization lead to representations of a kind of 'super-periphery' in the Western Balkans,[39] while the Baltic states of Latvia and Lithuania are seen more as semi-periphery, with Estonia striving towards the status of a Nordic core economy.[40] Clearly, emigration trends play a role in representations of both cores and peripheries. But emigration, together with low birth rates, go hand in hand with another peripheralization process, namely depopulation, whose effects are generally less thoroughly researched than those attributed more directly to migration. The representation that such modes of migration imply is that these countries are not doing well economically, experiencing brain drain and youth drain, which further lead to depopulation and rapid ageing.

Depopulation and 'brain drain'

In the Baltic discourses, but not in Estonia, depopulation trumps 'brain drain'. Both are present in Western Balkan discourses, but here there is more emphasis on 'brain drain'.

For comparative reasons, population trends in both regions are presented in Table 1. This shows widespread and notable population decline in most of the countries listed. Slight increases of population are recorded only in Montenegro and the Republic of Northern Macedonia for the 10 years 2005–2015. Moreover, much of the emigration took place before these years and thus the most recent decade simply continues the declining trend already apparent in the early 1990s. In relative terms, there has been at least a 12–16% population decline in Latvia and Lithuania due to emigration since the 1990s. But apart from statistical representation, depopulation is 'felt' across much of the Baltic region, where emptying towns and villages are observed away from the national capitals and other main cities.[41] Hence peripheralization has a 'structure of feeling', which goes above and beyond the picture yielded by scientific portrayals and periodic census data, which can be several years out of date.

According to Eurostat data published in 2018, population decline continues rapidly and the Baltic and Balkan countries are leaders in this trend: the steepest decrease was in Lithuania at 13.8 per 1000 residents per year, followed by Croatia at 11.8, and Latvia at 8.1. All figures are for the year 2017. And yet Estonia's population increased by 2.7 per 1000

Table 1. Population: Western Balkans and Baltics (millions).

Country	2005	2015
Albania	3.01	2.88
Bosnia and Herzegovina	3.78	3.54
Croatia	4.44	4.20
Kosovo	2.00	1.80
Montenegro	0.613	0.622
The Republic of Northern Macedonia	2.04	2.08
Serbia	7.5	7.10
Estonia	1.36	1.32
Latvia	2.30	1.98
Lithuania	3.41	2.91

Sources: European Commission key figures on enlargement countries, 2017; Statistical offices of Estonia, Latvia and Lithuania.

residents. However, Estonia's specific situation is related to 'commuter migrants' between Estonia and nearby Finland, who are not always counted into the statistics of emigration. While on average the EU population grew by 2.1 per 1000 in 2017, decline was observed in many southern and south-eastern member states: Bulgaria (−7.3 per 1000 residents), Romania (−6.2), Greece (−2.7), Hungary (−2.0), Portugal (−1.8) and Italy (−1.7).[42]

Depopulation is generally linked to weak development, especially in Western Balkan discourses, as the latest surveys of graduate migration and intentions to leave show.[43] Suggestions to boost jobs in order to tackle structural brain drain due to lack of opportunities are highly visible in Balkan representations of this issue.[44] They are less apparent in the Baltics despite broadly similar structural conditions. Brain drain has been an ongoing concern for the Balkans for several decades now, while the broader emigration trend changed around the 2000s when more low-skilled and older people emigrated.[45]

In the Baltics, more focus is on the issue of skills mismatch and the record levels of emigration; for Lithuania, this is related to over-education in transferable skills of Lithuania's workforce.[46] 'Brain drain' as such was less evident in the Baltic scientific discourse on migration, overtaken instead by skills mismatch and the deskilling of Baltic migrants already working abroad.[47] During the economic crisis in the Baltics, a strongly neoliberal approach was taken, representing emigration as a rational choice: to put it simply, if there are reduced chances to get a job at home, then go abroad. However, when analysing 'brain drain' through educational differences among emigrants and those who stayed put, Estonian migration scholars came to the conclusion that there was no significant brain drain after the EU enlargement.[48] However, differences have been found in relation to return migration intentions and integration levels in the host countries; ethnic Russians tend to express lower intention to return compared to ethnic Estonians.[49] Similar trends have also been found in Latvia.[50]

Transit migration: the very Western Balkan 'problem'

For some time the dominant representations within the Balkan field of migration research were concerned with wars, nationalism, displacement and geographical mobility rights.[51] This creates a particular mode of peripheralization: the Western Balkans as a periphery outside the 'club' of fortress Europe, and the construction of regional stability needs in terms of migration.[52] Dominated by discursive representations of 'illegal migration' and conditionality for institutions in the region to deal with it,[53] *transit migration* poses a special additional challenge to the above-discussed processes of East–West migration, depopulation and 'brain drain'. While emigration has clearly been associated with weak development in the Western Balkan region, discourses of 'crisis' in the Western Balkans have seen particularly linked to 'refugee crises': in the 1980s and 1990s as 'exodus' from the Balkans, and in 2015 as a transit route via the Western Balkans.[54] It is estimated that over a million migrants and asylum seekers transited the Western Balkan route in 2015 and 2016.[55] The flanking peripheralization effect is that these processes lead towards restrictive migration policies and requirements to increase capacities to deal with migration in Western Balkan countries which have yet to receive any specific time frame for potential EU accession. Moreover, this mode of peripheralization—constructed as a strategic geopolitical location for migration/refugee routes along with considerable ongoing emigration and weak development —gives rise to discourses on the lack of capacity and even willingness of the Western Balkan countries to deal with requirements to process asylum applications.[56]

Conversely, a major decline in irregular migration implies the effective willingness to control migration along with the capacity to fight organized crime related to some irregular migration processes, and thus move away from a periphery-as-transit migration discourse.[57] Prevention and control of irregular migration were among the pre-accession objectives in the Western Balkans in the late 2000s.[58] Migration control is also a factor which leads towards the EU (or away from it if unsuccessful); hence peripheralization is constructed as a lack of 'Europeanization' trends.[59]

Return migration

While in the Balkans emigration was possible throughout the twentieth century, except Albania, mass labour emigration from the Baltics began again only in the 1990s, accelerating sharply after joining the EU in 2004. While some 'guestworkers' from Balkans did return in the 1970s and 1980s, as already mentioned, some of them left again, unwilling or unable to readjust to socialist rule. But return also took place among more recent migrants, including refugees from Bosnia.[60] Return in later life, post-retirement, is also a trend, if not yet in large numbers in the Balkans.

But returns to peripheries or semi-peripheries are more complex than the theories which take the core–periphery imaginary as a standard of migration dynamics have often imagined. The Balkans, again, are a good example. During the severe Greek economic crisis, return migration to Albania also occurred, on a larger scale than from Italy or the 'core' economies in Europe or elsewhere that Albanians had emigrated.[61] Greece, along with Italy, hosts most of the Albanian migrants, partly related to geographic proximity, but many Albanian men lost their jobs during the crisis in Greece. Many of them returned to Albania to explore work opportunities there.[62] Crisis-induced returns from Greece to Albania signify the necessity for a more nuanced analysis of how migration and return migration flows are embedded within core–periphery dynamics. In this particular case, if we class Albania as periphery and Greece as semi-periphery, then migration from the former to the latter creates a fragile dependency which is exposed when the semi-peripheral country stumbles into a deep crisis.[63]

It has been noted that the countries of Eastern Europe seem to be slower than other European countries not only in return trends but also in attracting immigration flows.[64] Accordingly, peripheralization, seen from the migration perspective, persists over time in these regions. The exception is Estonia, where immigration has overtaken emigration in 2017. Return to the Baltics is still rarely discussed in scientific texts on migration. When the issue of return has been mentioned in international publications, the focus has been on high aspirations to return and high selectivity towards young and entrepreneurial talent in Lithuania[65] and Estonia,[66] with somewhat more cautious discourses on the actual potential to transfer skills back home in Latvia.[67] Quality of life in the Baltics upon return can indeed be somewhat higher, but mainly for certain groups, usually the higher skilled and those working in high-demand and well-paid professions.[68]

The young and highly skilled aspire to return to Serbia too, as Prelec's research demonstrates in this special issue.[69] However, they are highly critical towards nationalistic stances in their country of origin, which leaves a gap between their high aspirations and the actual decisions to remain abroad and engage in home politics through transnational links, such as external voting.

The transnational migration perspective, where permanent return is not envisaged as the primary option, is gradually emerging in both the Balkans[70] and the Baltics.[71] This approach can also be seen in a variety of diaspora initiatives in both regions. There is no lack of initiatives for so-called global leaders,[72] diaspora youth, return migration of talented individuals and work opportunities for the highly skilled.[73] Civic innovation incubators, global diaspora strategies,[74] high-level political representation—all these are similar in all countries but still with rather minimal actual results to report.

The impact of the emigration of skilled people on development in the Western Balkans remains to be explored from varied perspectives, including transnational ties, which may include, but not necessarily result in, return migration.[75] Yet there are alarming voices which maintain that 'knowledge emigration' could have more adverse future effects on development than overall emigration. On the other hand, talent migration, if there are strong diaspora and return incentives, can even be beneficial if ties with the country of origin are maintained.[76]

Conclusions

Firstly, the construction of emigration, represented scientifically as the East–West model for several decades, reinforces the representation of peripheries. As such, this process is a social construction of peripheralization. This is not to say that there are no 'real facts' about emigration—this was not the main point of the paper. My argument is that socially constructed representations contribute to the ways in which we perceive, comprehend and continue to represent peripheralization.

Secondly, the representation of depopulation adds even more to the perception of continuous peripheralization in these regions, along with a narrowed future development potential due to brain drain. The latter, however, is represented in more varied ways in the Balkans and Baltics. Estonia stands out as an example where brain drain is actually not perceived as an emigration-induced problem.

These seemingly straightforward findings require a theoretical discussion as well. Peripheralization, unlike the simple definition of peripheries as territorial units ranked lower than cores, is a process and creation of perception. With respect to regional peripheralization, it is obvious how representation matters. Both the Balkans and the Baltics have been historical peripheries within empires and in socialist unions during the twentieth century, with the partial exception of Serbia which, as part of Yugoslavia, could be considered as its centre of power. As contemporary peripheries these regions can be relegated vis-à-vis the European Union with the Baltics being members since 2004, while most of the Western Balkan countries have not yet received a concrete timetable for joining the EU (with exception of Croatia, which joined smoothly with rather little public notice in 2013). What is important in terms of peripheralization as a process is how some countries, individually and not as regions, are represented as moving away from a periphery or semi-periphery, as with the case of Estonia. Paasi's questions of 'who and what constructs regions, through what associations/networks and for what purposes?'[77] are central here, using migration as a lens through which we can think about peripheralization.

The response from my findings is that making a threshold turn when net immigration replaces net emigration is a milestone event in the ways through which migration matters in the representation of countries. Such predictions that successful fulfilling of the required conditions for EU accession would make countries attractive to immigrants are widespread. Yet this has not been the case in the Baltics so far, and even in the case of Estonia it is unclear whether EU accession as such makes the country more attractive for immigration. The opposite is clearer: nationalistic politics deter both return migration and immigration. Staunch opposition to immigration and emphasis on diaspora policies and the return of 'our people'—co-ethnics—reveal a mode of peripheralization which stems from within countries rather than being externally constructed.

As states, the Western Balkans and the Baltics reveal considerable diversity, individualization and the deliberate choosing of a regional affinity discursively; again with Estonia as a stand-out example, representing itself the most vocally as a Nordic, rather than a Baltic state. But there are also similarities, especially due to the geopolitical positioning of these regions as 'margins' of Europe. For the Western Balkans it is especially strong through the mode of transit migration which represents the region as a buffer outside the 'fortress' EU and yet with limited capacities and 'willingness' to deal with asylum procedures. Recent transit migration via the Western Balkans represents the case of a specific challenge.

I want to conclude with two perspectives which could make a difference in reimagining peripheralization processes through migration. Firstly, the representation of modes of migration takes us beyond current understandings that the periphery as such can remain a relatively stable scientific concept, even if 'never fully complete and stable'.[78] The migration perspective not only adds a factor on how peripheralization can be measured and interpreted; moreover, it adds a dynamic element and destabilizes the otherwise relatively stable thinking on spatial units. And secondly, transnational and relational perspectives necessitate that studies of migration to and from relatively small states and territories be placed on equal footing with those of core states and core languages. When theories and research emerge from such places, and not only about these places, we may reimagine cores and peripheries in a different light.

Notes

1. For a rare exception see R. King, J. Vullnetari, A. Lulle and E. Cela, 'Contrasts in ageing and agency in family migratory contexts: a comparison of Albanian and Latvian older migrants', in E. Palenga-Möllenbeck and M. Kilkey (eds), *Family Life in an Age of Migration and Mobility: Global Perspectives Through the Life Course*, Palgrave Macmillan, Basingstoke, 2016, pp. 261–286.
2. A. Paasi, 'Regions are social constructs, but who or what "constructs" them? Agency in question', *Environment and Planning A*, 42(10), 2010, pp. 2296–2301.
3. Ibid., p. 2298.
4. L. Wolff, *Inventing Eastern Europe: The Map of Civilization on the Mind of the Enlightenment*, Stanford University Press, Stanford, CA, 1994.
5. See R. King, R. Skeldon and J. Vullnetari, *Internal and International Migration: Bridging the Theoretical Divide*, University of Sussex, Sussex Centre for Migration Research, Working Paper No. 52, 2008.
6. Paasi, op. cit., pp. 2297, 2301.
7. Z. Krisjane and M. Berzins, 'Post-socialist urban trends: new patterns and motivations for migration in the suburban areas of Rīga, Latvia', *Urban Studies*, 49(2), 2012, pp. 289–306.
8. See T. Lang, S. Henn, W. Sbignev and K. Ehrlich, *Understanding Geographies of Polarization and Peripheralization: Perspectives from Central and Eastern Europe and Beyond*, Palgrave Macmillan, Basingstoke, 2015, p. 12.
9. W. Bartlett and I. Prica, 'Debt in the super-periphery: the case of the Western Balkans', *Third World Thematics: A TWQ Journal*, 2018, DOI: 10.1080/23802014.2017.1438850.
10. See J. P. Cassarino, 'Theorising return migration: the conceptual approach to return migrants revisited', *International Journal on Multicultural Societies*, 6(2), 2004, pp. 253–279.
11. For the most complete account of European core–periphery dynamics in the context of migration see R. King, 'Migration in Southern Europe—a centre–periphery dynamic?', in M. Baumeister and R. Sala (eds), *Southern Europe? Italy, Spain, Portugal and Greece from the 1950s to the Present* day, Campus Verlag, Frankfurt, 2015, pp. 139–169. See also R. King, A. Lulle, V. Parutis and M. Saar, 'From peripheral region to escalator region in Europe: young Baltic graduates in London', *European Urban and Regional Studies*, 25(3), 2018, pp. 284–299.
12. A. G. Frank, *Capitalism and Underdevelopment in Latin America*, Monthly Review Press, New York, 1969.
13. I. Wallerstein, *The Modern World-System*, Academic Press, New York, 1974.
14. D. Seers, B. Schaffer and M.-L. Kiljunen (eds), *Underdeveloped Europe: Studies in Core–Periphery Relations*, Harvester Press, Hassocks, 1979.
15. P. Selwyn, 'Some thoughts on cores and peripheries', in D. Seers, B. Schaffer and M.-L. Kiljunen (eds), *Underdeveloped Europe: Studies in Core–Periphery Relations*, Harvester Press, Hassocks, 1979, pp. 35–43.
16. S. Beetz, S. Huning and T. Plieninger, 'Landscapes of peripheralization in North-Eastern Germany's countryside: new challenges for planning theory and practice', *International Planning Studies*, 13(4), 2008, p. 305.
17. M. Kuus, 'Places of lower rank: margins in conversations', *Political Geography*, 37(A1–2), 2013, pp. 30–32.
18. For a recent detailed study of the Albanian brain drain see I. Gedeshi and R. King, *Research Study into Brain Gain: Reversing Brain Drain with the Albanian Scientific Diaspora*, Centre for Economic and Social Studies, Tirana, report prepared for UNDP.
19. D. J. Galbreath and J. W. Lamoreaux, 'Punching above your weight? The Baltic States as "small states" and the European neighbourhood policy?', in Z. Ozolina and N. Muiznieks (eds), *Uncertain Transformations. New Domestic and International Challenges*, LU Akdemiskais Apgads, Riga, 2007, pp. 21–33; M. Todorova, *Imagining the Balkans*, Oxford University Press, Oxford, 2009.

20. M. Saar, *The Answers You Seek Will Never be Found at Home: Reflexivity, Biographical Narratives and Lifestyle Migration among Highly Skilled Estonians*, PhD dissertation, Södertörns Högskola, Huddinge, 2017.
21. Paasi, op. cit.; see also D. Light and C. Young, 'European Union enlargement, post-accession migration and imaginative geographies of the "New Europe": Media discourses in Romania and the United Kingdom', *Journal of Cultural Geography*, 26(3), 2009, pp. 281–303.
22. See H. Bauder and G. Gilbert, 'Representations of labour migration in Guatemalan and American media', *ACME: An International E-Journal for Critical Geographies*, 8(2), 2009, p. 283; also H. Bauder, 'Landscape and scale in media representations: the construction of offshore farm labour in Ontario, Canada', *Cultural Geographies*, 12(1), 2005, pp. 41–58.
23. D. Ljuboja, *Migration Processes in the Balkans in the 20th and 21st Centuries*, 2015 <http://www.cirsd.org/en/young-contributors/migration-processes-in-the-balkans-in-the-20th-and-21st-centuries> (accessed 15 July 2018).
24. Ibid.
25. R. King, 'Albania as a laboratory for the study of migration and development', *Journal of Southern Europe and the Balkans*, 7(2), 2005, pp. 133–155.
26. For more details see the collection of papers in J. Vullnetari and R. King (eds), 'Twenty Years of Albanian Migration', *Journal of Balkan and Near Eastern Studies*, 13(3), 2011, pp. 269–356.
27. R. King, E. Uruçi and J. Vullnetari, 'Albanian migration and its effects in comparative perspective', *Journal of Balkan and Near Eastern Studies*, 13(3), 2011, pp. 269–286.
28. J. Kielyte and D. Kancs, 'Migration in the enlarged European Union: empirical evidence for labour mobility in the Baltic states', *Journal of Baltic Studies*, 33(3), 2002, pp. 259–279.
29. Latvia Population Census, 2011 <https://www.csb.gov.lv/en/statistics/statistics-by-theme/population/census>; Lithuania Population Census, 2011 <https://osp.stat.gov.lt/en/gyven toju-ir-bustu-surasymai1>.
30. G. Engbersen and J. Jansen, 'Emigration from the Baltic States: economic impact and policy implications', in OECD, *Coping with Emigration in Baltic and East European Countries*, OECD, Paris, 2013, p. 13.
31. See F. Keereman and I. Szekely (eds), *Five Years of an Enlarged EU: A Positive Sum Game*, Springer-Verlag, Berlin, 2010, pp. 63–94.
32. A. Lulle, 'Baltic triplets? Out-migration and responses to crisis', *Regional Formation and Development Studies*, 3(11), 2013, pp. 135–145; also A. Zaiceva and K.F. Zimmermann, 'Returning home at times of trouble? Return migration of EU enlargement migrants during the crisis', in M. Kahanec and K. F. Zimmermann (eds), *Labor Migration, EU Enlargement, and the Great Recession*, Springer, Berlin, 2016, pp. 397–418.
33. J. Reineck, *The Past as Refuge: Gender, Migration, and Ideology among Kosova Albanians*, PhD dissertation, University of California, Berkeley, 1991.
34. *Diaspora as a Driving Force for Development in Kosovo: Myth or Reality?*, Forum for Democratic Initiatives, Gjakove, Kosovo, 2009 <https://www.eda.admin.ch/dam/deza/en/documents/laender/resource-183043_EN.pdf> (accessed 21 July 2018); J. Möllers, W. Meyer, S. Xhema, D. Traikova and G. Buchenrieder, 'Cognitive constructs and the intention to remit', *Journal of Development Studies*, 51(10), 2015, pp. 1341–1357.
35. See *Migration and Remittances Factbook*, World Bank, Washington, DC, 2011, pp. 54, 180.
36. M. Kupiszewski, A. Kicinger, D. Kupiszewska and E. F. Flinterman, *Labour Migration Patterns, Policies and Migration Propensity in the Western Balkans*, International Organization for Migration, Budapest, 2009, pp. 12–13.
37. F. W. Carter, 'Ethnicity as a cause of migration in Eastern Europe', *GeoJournal*, 30(3), 1993, pp. 241–248.
38. B. Balli, 'From communist past to capitalist present: labour market experiences of Albanian immigrants in Greece', *Journal of Balkan and Near Eastern Studies*, 13(3), 2011, pp. 323–337.

39. W. Bartlett, 'Economic development in the European super-periphery: evidence from the Western Balkans', *Economic Annals*, 54(181), 2009, pp. 21–44.
40. King, Lulle, *et al.*, op. cit.
41. T. Lukic, R. Stojsavljevic, B. Durdev, I. Nad and B. Dercan, 'Depopulation in the Western Balkan countries', *European Journal of Geography*, 3(2), 2012, pp. 6–23; D. Dzenovska, *Aizbraukšana un tukšums Latvijas laukos: Starp zudušām un iespējamām nākotnēm* (Emigration and emptiness. Between lost and possible futures), Biznesa Augstskola Turība, Riga, 2012.
42. Eurostat, population trends <http://ec.europa.eu/eurostat/web/products-eurostat-news/-/WDN-20180711-1?inheritRedirect=true&redirect=%2Feurostat%2F> (accessed 21 July 2018).
43. I. Gedeshi, *The Global Crises and Emigration—Monitoring a Key Transmission Channel to the Albanian Economy*, UNDP/IOM/WB, Tirana, Albania, 2010.
44. D. Kovtun, A. Mayer Cirkel, Z. Murgasova, D. Smith and S. Tambunlertchai, *Boosting Job Growth in the Western Balkans*, IOM, 2014 <https://www.imf.org/en/Publications/WP/Issues/2016/12/31/Boosting-Job-Growth-in-the-Western-Balkans-41284> (accessed 21 July 2018).
45. E. Germenji and L. Milo, 'Return and labour status at home: evidence from returnees in Albania', *Southeast European and Black Sea Studies*, 9(4), 2009, pp. 497–517.
46. Z. Norkus, *On Baltic Slovenia and Adriatic Lithuania: A Qualitative Comparative Analysis of Patterns in Post-Communist Transformation*, CEU Press, Budapest, 2012.
47. M. Hazans and K. Philips, 'The post-enlargement migration experience in the Baltic labor markets', in M. Kahanec and K. F. Zimmermann (eds), *EU Labor Markets after Post-Enlargement Migration*, Springer, Berlin and Heidelberg, 2010, pp. 255–304; R. Eamets, 'Employment, labour market flexibility and the economic crisis in the Baltic states', in *Estonian Human Development Report*, Eesti Koostoo Kogu, Tallinn, 2011, p. 74.
48. K. Anniste, T. Tammaru, E. Pungas and T. Paas, *Emigration after EU Enlargement: Was There a Brain Drain Effect in the Case of Estonia?*, Working Paper No. 87, University of Tartu, 2012.
49. K. Anniste and T. Tammaru, 'Ethnic differences in integration levels and return migration intentions: a study of Estonian migrants in Finland', *Demographic Research*, 30(1), 2014, pp. 377–412.
50. S. Aptekar, 'Contexts of exit in the migration of Russian speakers from the Baltic countries to Ireland', *Ethnicities*, 9(4), 2009, pp. 507–526.
51. M. H. Yavuz and I. Blumi, *War and Nationalism: The Balkan Wars 1912–1913, and their Sociopolitical Implications*, University of Utah Press, Salt Lake City, Utah, 2013.
52. A. Geddes and A. Taylor, 'In the shadow of fortress Europe? Impacts of European migration governance on Slovenia, Croatia and Macedonia', *Journal of Ethnic and Migration Studies*, 42(4), 2016, pp. 587–605; N. Tzifakis, 'EU's region-building and boundary-drawing policies: the European approach to the Southern Mediterranean and the Western Balkans', *Journal of Southern Europe and the Balkans*, 9(1), 2007, pp. 47–64.
53. F. Trauner, 'From membership conditionality to policy conditionality: EU external governance in South Eastern Europe', *Journal of European Public Policy*, 16(5), 2009, pp. 774–790.
54. J. Sardelic, *From Temporary Protection to Transit Migration: Responses to Refugee Crises along the Western Balkan Route*, European University Institute, Florence, RSCAS Working paper 35, 2017.
55. J. Arsenijević, E. Schillberg, A. Ponthieu, L. Malvisi, W. A. Elrahman Ahmed, S. Argenziano, F. Zamatto, S. Burroughs, N.e Severy, C. Hebting, B. de Vingne, A. D. Harries and R. Zacharia, 'A crisis of protection and safe passage: violence experienced by migrants/refugees travelling along the Western Balkan corridor to Northern Europe', *Conflict and Health*, 11(6), 2017, pp. 1–9.

56. N. Kogovšek Šalamon, 'Asylum systems in the Western Balkan countries: current issues', *International Migration*, 54(6), 2016, pp. 151–163.
57. M. Baldwin-Edwards, 'Sustainable development and emigration: the contemporary Balkans and the European Union', *SEER: Journal for Labour and Social Affairs in Eastern Europe*, 7(1), 2004, pp. 9–14.
58. F. Trauner, *EU Justice and Home Affairs Strategy in the Western Balkans: Conflicting Objectives in the Pre-Accession Strategy*, CEPS Working Document No. 259, 2007 <http://aei.pitt.edu/7383/2/7383.pdf> (accessed 21 July 2018).
59. S. Kacarska, *Europeanisation through Mobility: Visa Liberalisation and Citizenship Regimes in the Western Balkans*, CITSEE Working Paper 11, 2011.
60. A. H. Stefansson, 'Homes in the making: property restitution, refugee return, and senses of belonging in a post-war Bosnian town', *International Migration*, 44 (3), 2006, pp. 115–139.
61. R. Panagiotou, 'The Greek crisis as a crisis of EU enlargement: how will the Western Balkans be affected?', *Southeast European and Black Sea Studies*, 13(1), 2013, pp. 89–104.
62. E. Caro, '"I am an emigrant in my own country": a profile of Albanian returnees from Greece', *Journal of Science and Higher Education Research*, 1(1), 2016, pp. 9–25.
63. D. Bechev, *The Periphery of the Periphery: The Western Balkans and the Euro Crisis*, European Council of Foreign Relations, 2012 <https://www.ecfr.eu/publications/summary/the_periphery_of_the_periphery_the_western_balkans_and_the_euro_crisis> (accessed 21 July 2018).
64. M. Okolski, *Europe in Movement: Migration from/to Central and Eastern Europe*, CMR Working Papers 22/80, 2007.
65. E. Barcevičius, 'How successful are highly qualified return migrants in the Lithuanian labour market?', *International Migration*, 54(3), 2015, pp. 35–47.
66. Saar, op. cit.
67. R. King, A. Lulle and L. Buzinska, 'Beyond remittances: knowledge transfer among highly educated Latvian youth abroad', *Sociology of Development*, 2(2), 2016, pp. 183–203.
68. Saar, op. cit.
69. T. Prelec, 'The Serbian diaspora's political views: a study on the 2017 Serbian presidential election', *Journal of Balkan and Near Eastern Studies*, this issue.
70. See for instance S. Bozic, 'Is there a Croatian diaspora in Europe? From "Gastarbeiters" to transmigrants and ethnics', *Croatian Studies Review*, 8, 2012, pp. 117–134; S. Bozic and S. Kuti, 'New international borders old social spaces: transnational migrant networks across the boundaries of post-socialist Croatia', *Österreichische Zeitschrift für Soziologie / Austrian Journal of Sociology*, 41(4), 2016, pp. 409–426.
71. A. Lulle and L. Buzinska, 'Between a "student abroad" and "being from Latvia": inequalities of access, prestige, and foreign-earned cultural capital', *Journal of Ethnic and Migration Studies*, 43(8), 2017, pp. 1362–1378.
72. *Programmes for Lithuanian Emigrants* <http://www.enterpriselithuania.com/en/> (accessed 21 July 2018).
73. 'Talents Back Home!' <http://www.workinestonia.ee> (accessed 21 July 2018).
74. Croatian Diaspora Congress—Zagreb 2014 <http://www.domovod.info/forum.php> (accessed 21 July 2018).
75. E. Germenji and L. Milo, 'Migration of the skilled from Albania: Brain drain or brain gain? ', *Journal of Balkan and Near Eastern Studies*, 13(3), 2011, pp. 339–356; and more broadly R. King, M. Frykman and J. Vullnetari (eds), *Migration, Transnationalism and Development in South-East Europe and the Black Sea Region*, Routledge, London, 2017.
76. D. Kancs and J. Kielyte, 'Education in the East, emigrating to the West?', *European Review*, 18(2), 2010, pp. 133–154.
77. Paasi, op. cit.
78. Ibid., p. 2229.

Disclosure statement

No potential conflict of interest was reported by the author.

Refugees at the Gates of the EU: Civic Initiatives and Grassroots Responses to the Refugee Crisis along the Western Balkans Route

Chiara Milan

ABSTRACT

The countries of former Yugoslavia have a long history of dealing with refugees. However, what had been termed in the media and in public discourse as the '2015–2016 European migrant crisis' represented a phenomenon of unprecedented scale for the Western Balkan states. For the first time since the break-up of Yugoslavia, these countries found themselves handling a steep increase in arrivals of migrants who fled from the Middle East and Africa and crossed the Western Balkans migratory route in an attempt to reach Northern and Central Europe. To compensate for the lack of action on the part of their governments, from summer 2015 a number of individuals and domestic groups mobilized with the purpose of providing humanitarian assistance to migrants crossing their territory. Based on a series of in-depth interviews with key informants, this article explores the multiple solidarity initiatives organized at the local and transnational level along the Western Balkans route during the 2015–2016 migrant crisis, taking Serbia and Macedonia as case studies. Specifically, the study examines the extent to which emotional resources and discursive opportunities increased the prospects for altruistic mobilization in a context characterized by a low level of grassroots civic activism.

Introduction

The former Yugoslav states have traditionally been considered countries of emigration. In particular from 1975 onwards, internal migratory movements drove people from southern Yugoslavia to become guest-workers in Slovenia. Later on, the aftermath of the 1992–1995 conflicts pushed thousands of citizens of the former Yugoslav republics to migrate to Northern European countries like Germany, Sweden and Austria in search of jobs and safety. As a consequence of the war, the number of civilians forced to flee their homes to seek a safe haven in neighbouring towns and states also increased dramatically. The Office of the United Nations High Commissioner for Refugees (UNHCR) defines them as internally displaced people (IDPs), as these individuals migrated within the borders of their own country and remained under the protection of their government, even though the latter was the reason for their displacement.[1] As of late 2014, the number of IDPs and refugees estimated still to be present in Bosnia-Herzegovina was around 100,000;[2] in Serbia that figure amounted to 97,000 in May 2015.[3] Over summer 2015, the territories of the former Yugoslav countries were in the media limelight for being crossed in record-breaking numbers by people fleeing from the Middle East and Africa seeking protection in the European Union. In an attempt to reach Northern and Central Europe, migrants travelled along the Western Balkans route, one of the most popular migratory paths leading to the EU. The European Commission estimated that nearly 700,000 people had travelled across the route by the end of October 2015, the figure reaching 760,000 by the end of December of the same year.[4] What the media termed the '2015–2016 European migrant[5] crisis' represented a challenge of unprecedented scale for countries still dealing with internal displacement from the conflicts of the 1990s, and affected for the first time in their post-war history by such a massive influx of people.

The growth in the migration flow along the European trail had profound effects on the former Yugoslav countries, which found themselves on one of the main refugee routes. Holding a crucial geographical position at the borders of the EU, the Western Balkan states turned into countries of transit for migrants first passing through Turkey and Greece. While, on the one hand, the refugee crisis is said to have put the countries of former Yugoslavia 'back on the political map of Europe',[6] and 'the strategic importance of the region for the EU's stability and security' has been stressed,[7] on the other hand it has negatively impacted the relations among them, accentuating the lack of cooperation among policy-makers in the region.[8] The high number of arrivals at the EU borders represented a challenge for state institutions, which proved ill-prepared to face it, as well as to provide state protection and appropriate reception structures. At the transnational level, tensions between neighbourhood governments escalated during summer 2015, the peak of the migration crisis. A case in point is the controversy between Serbia and Croatia, which culminated in the closure of the Croatian border to Serbian goods,[9] and into the Croatian veto on Serbia opening chapters 23 and 24 of its EU accession process.[10]

Confronted with the complex effects posed by the significant number of refugees, at the national level, government officials reacted unevenly. A shifting attitude characterized the response of the Macedonian authorities, the first non-EU country at the southern EU border.[11] At first, in early 2015, government officials prohibited migrants from boarding public transport to cross the country's territory, and periodically shut the Greek border crossings in an attempt to stem the flow of people. A couple of months after permitting migrants to legally use the country's public transportation system, in August 2015 Macedonian authorities sent the riot police to the Greek border to push migrants back with truncheons and riot shields.[12] The government repeatedly blamed the Greek authorities for being incapable of dealing with the influx of people, and, as a response, started to erect wire fences along the Greek border in February 2016. Serbia, headed by the then Prime Minister (PM) Aleksandar Vučic,[13] at first adopted an open-door and open-borders policy vis-à-vis migrant populations, which later changed following the closure of the Western Balkans route. The Slovenian and Croatian governments, affected by the influx of migrants after the sealing of the Hungarian–Serbian frontier in September 2015, resorted mostly to conventional border enforcement measures. However, during the period of semi-official establishment of the

Western Balkans corridor both countries provided free-of-charge transport to facilitate transit migration towards Central and Northern Europe.[14] Later, though, both countries placed restrictions on the nationalities allowed to cross their frontiers. In February 2016, Slovenia introduced a daily cap for migrants admitted into the country, before closing off its borders. Croatia, Serbia and Macedonia followed suit. Eventually, a series of meetings headed by the Austrian government sanctioned the closure of the Western Balkan corridor on 9 March 2016. After the shutdown, Slovenia and Croatia reintroduced the Schengen Border Code[15] and started to expel migrants from their territories.

The European migratory crisis also represented a watershed at the societal level, for it spurred a civil response both locally and transnationally. Contrary to the divisive behaviour of political leaders, a part of civil society responded to migrant arrivals with solidarity actions and humanitarian initiatives. To compensate for the poor responsiveness of their governments, a number of individuals mobilized to provide immediate relief to migrants in transit during their journey towards the EU. As in other parts of Europe,[16] the type of civic response varied over time and space, ranging from advocacy-oriented types of intervention, stop-gap self-organized initiatives, to small-scale demonstrative types of actions at the border crossings. Independently run, volunteer-based organizations provided an immediate response to the increased arrivals. In several cases, solidarity groups reacted more immediately and at a faster pace than governments. Active at the border crossings and in the main cities the migrants crossed, volunteers and grassroots groups brought about an alternative system of humanitarian assistance. With the unfolding of the crisis, and the increasingly restrictive measures imposed by their governments, solidarity groups found themselves progressively prevented from providing assistance to migrants. Following the decision of governments to intervene and take over the task of assisting migrants, state-run registration camps underwent a process of securitization and professionalization.[17] As a consequence, informal solidarity groups found themselves excluded from accessing the transit camps hosting asylum seekers unless in the framework of large official organizations.

In a context that holds a weak record of grassroots activism, the emergence of a solidarity movement stands as a phenomenon that needs to be investigated further and from different angles. While, on the one hand, solidarity actors along the Western Balkans route mobilized at the local level, on the other they developed cross-national bonds with groups emerging with a similar purpose in other countries along the route. Interestingly, several individuals engaged in pro-refugee activism had not previously been involved in solidarity initiatives, and many had no prior record of political activism. By contrast, other groups had already been active in the territory for some years.

Against this background, this article explores the conditions that favoured the emergence and unfolding of pro-refugee solidarity initiatives in two countries located at the outskirts of the EU, both holding a weak and recent record of grassroots activism. Specifically, I contend that a combination of two elements in particular increased the prospects for solidarity initiatives to thrive, compensating for the weak ties amongst activists and the limited financial and organizational resources, deemed in social movement studies necessary preconditions for collective action. First, the emotional involvement of the local population with refugees in transit, often grounded on a personal experience of displacement or uprooting. Second, the narratives of government officials, which framed the migratory movement through a discourse of transience and exceptionality, for which the EU had to be deemed responsible, while citizens were to be praised for their spontaneous, altruistic response. Essentially, with this study I wish to demonstrate that, in an emergency context, solidarity movements[18] can emerge and blossom notwithstanding limited financial and organizational capacities, if there is an emotional involvement with the groups of individuals concerned and if discursive opportunities are favourable. To demonstrate that, I investigate the accounts of the migrant crisis given by movement actors in Serbia and Macedonia, two countries located at the outskirts of the EU, and heavily involved in the passage of migrants during the 2015–2016 migratory crisis. These accounts are aimed at reconstructing how solidarity actors motivated their engagement, and at explaining their decision to become involved in pro-refugee activism. To that end, I cast light on the discourses of government officials, in order to understand the extent to which the societal and political environment in which solidarity initiatives developed influenced its emergence and unfolding.

In what follows, I first outline the evolving scenario of the Western Balkans route and the semi-official establishment of the corridor, together with the grassroots response to the migrant flow in Macedonia and Serbia, the two countries on which the study focuses. Next, I explore in detail the two factors that I contend contributed to fostering altruistic mobilizations along the migratory path—emotional involvement and discursive opening. Finally, I draw some conclusion by summarizing the findings of the study.

Methodological note

The article relies on 15 in-depth qualitative interviews with key informants (activists, NGO practitioners and volunteers, and journalists) engaged in solidarity initiatives with migrants during the period lasting from spring 2015 to summer 2016,[19] and on participant observation in Belgrade during late July 2016.[20] Of these 15 interviews, seven were conducted in Macedonia and eight in Serbia. The solidarity actors interviewed differ from an organizational point of view, and have different political backgrounds. The configuration includes established actors such as formal organizations involved with migration-related activism from the years before the 2015–2016 migrant crisis, which also engaged in lobbying campaigns towards their government to ask for changes in the country's asylum law (NGO Legis and Macedonian Young Lawyers Organization in Macedonia; Adventist Development and Relief Agency—ADRA—in Serbia). Additionally, I interviewed members of social movement collectives, which engaged in actions of protest and awareness-raising on the topic of migration, besides providing relief to refugees in need (i.e. Solidarnost in Macedonia and No Borders group in Serbia); humanitarian organizations and local associations which emerged during the migrant crisis (Help the Refugees in Macedonia, the Miksalište centre and the Info Park initiative in Serbia), also participated in by foreign activists and volunteers. Nevertheless, the demarcation amongst these groups is rather blurred, as in practice a significant number of them were involved in different types of activities, ranging from protest-oriented to solidarity-oriented initiatives. The study also draws on the analysis of statements of policy-makers, as well as content analysis of media reporting in international news outlets.

The Western Balkans corridor and the grassroots response to the 2015–2016 migrant crisis

When talking about the migratory trail through the Western Balkans region, a distinction is necessary between the Western Balkans route, an informal migratory path shaped over the years by movements of migrants striving to make their way across the borders of Europe, often with the help of human traffickers, and the Western Balkans corridor, which denotes a temporarily legalized, state-sponsored passageway. Between 2015 and early 2016, the latter allowed migrants to move from the Turkish shore and the Greek islands through the Balkans to Northern Europe. The corridor permitted hundreds of thousands of migrants hailing mostly from Syria, Afghanistan, Iran, Iraq, but also Pakistan, Eritrea, Somalia and Palestine, to pass across the previously closed borders of Macedonia, Serbia, Croatia, Slovenia and Austria, and to reach Central Europe in a relatively safe manner.[21] However, during the period in which the corridor remained open—approximately from September 2015 to March 2016—several states imposed nationality-based restrictions to entry in their territories. As of mid-November 2015, the passage through the corridor was denied to individuals not coming from Syria, Iraq and Afghanistan (the so-called SIA countries). In February 2016, a ban was put on Afghans wanting to cross the borders. Following the shutdown of the corridor in March 2016, only those in possession of a valid visa were allowed to cross through the former Yugoslav states, as the reintroduction of the Schengen Border Code was announced, marking the temporary reinstatement of border controls at the internal borders of the EU as of 8 March.[22]

Macedonia, the entry point of the corridor

Since 2015, the village of Gevgelija, in the proximity of the Greek border in southern Macedonia, represented the corridor's entry point. During the period lasting from September to November 2015, the number of new arrivals at the Macedonian southern border amounted to over 150,000 persons per month, which means more than 5000 new arrivals per day.[23] The figure dropped to around 1500 new persons per day in early 2016.[24] In total, between June 2015 and March 2016, the number of people who entered Macedonian territory declaring their intention to lodge an asylum claim in the EU amounted to almost 500,000.[25] The solidarity movement in the country sprang from a tragic event which occurred in April 2015, when a train struck 14 young migrants of Somali and Afghani origin who were walking along the railway tracks not far from the capital Skopje.[26] Irregular according to the country's asylum law, the migrants were forbidden by law to board public transport. Likely to be arrested and detained, they thus strove to reach the north of the country on foot, walking parallel to the train lines after sunset. The incident received high media coverage all over the country, to the extent that a significant number of interviewees reported it as marking an important turning point in raising public awareness on the refugee cause.[27] Following the incident, several individuals created a Facebook page called 'Help the refugees in Macedonia', which became a virtual platform to coordinate the provision and distribution of food, clothing and sanitary supplies to migrants on a daily basis.[28] Domestic NGOs mobilized in their support as well, reaching out to them on the streets, providing food supplies and offering legal support.[29] Following pressure from local and international NGOs, in June 2015 the national Law on Asylum and Temporary Protection was amended to allow migrants to use public transportation on the way to Serbia, at marked-up prices though.[30] In June 2015, the Macedonian parliament voted to give migrants a three-day visa that allowed them free movement across the country with the requirement that they left within 72 hours from their arrival in case they did not officially lodge an asylum claim. Allegedly unable to cope with the inflow of arrivals from Greece, in August 2015 the government declared a state of emergency on the southern and northern borders,[31] and deployed the army and security forces at the Greek border. On 18 November 2015, Macedonia introduced also a 'nationality screening programme' imposing nationality-based restrictions, which limited entry to the country exclusively to SIA nationals.[32] Ten days later, local authorities ordered the erection of a wire fence along the Greek border in an attempt to stem the influx of migrants from the country, before eventually closing the entry point at Gevgelija in early March 2016.

Serbia, the bottleneck

Since May 2015, Serbia witnessed a rise in the number of migrants transiting across its territory. The flow soared in July 2015, as a consequence of the Hungarian government's decision to close down its border with Serbia and to initiate construction of a fence along it (completed in mid-September 2015) to deter people from entering the country. Over summer 2015, thousands of migrants began to populate the parks[33] surrounding the train and bus stations of the capital, Belgrade, which became the main point where people on the move could find temporary refuge in tents. Local and foreign volunteers, organized in the 'Info Park' initiative, and operating from inside a former warehouse transformed into a collection/distribution centre and welcome point, named 'Miksalište', provided them daily with food, clothing and information.[34] Other informal refugee centres emerged in the capital, especially after the local authorities prevented migrants from pitching their tents in the parks. With the support of the local and international No Border group activists, several migrants sought shelter in another warehouse in the vicinity of Miksalište, which they squatted and named 'No Border Hostel'.[35] Eventually, both Miksalište and the No Border Hostel were dismantled in April 2016, officially to make room for the contested 'Belgrade Waterfront Project', a project aimed at the refurbishment of the capital's Sava Mala neighbourhood.[36] Besides self-organized initiatives, in Serbia domestic and foreign NGOs got involved in providing humanitarian assistance in state-run transit camps all over the country, also supporting local authorities in facilitating the transit of migrants along the corridor. During the opening of the Balkan corridor, volunteers and NGO practitioners took care of migrants upon their arrival in Serbia in cooperation with officials of the country's Commissariat for Refugees and Migration, a public organization in charge of dealing with the 'care, return and integration of refugees'[37]—as was the case of interviewees working or volunteering in the officially recognized transit centre of Preševo. As of late November 2015, access to the corridor was restricted only to SIA nationals.

Consequently, those individuals unable to prove Syrian, Iraqi or Afghan citizenship attempted to enter Serbia from across the Bulgarian frontier with the help of people smugglers. Serbia thus converted into the transit point for migrants travelling across both Macedonian and Bulgarian paths. Due to the shutdown of the corridor in early March 2016, Belgrade again became the bottleneck of the migration route, being a crucial nexus for all those striving to reach the Schengen space through Hungary or Croatia. Facing no other option than remaining in Serbia, migrants populated the parks of Belgrade until late July 2016, when the municipality ploughed and fenced the green areas. Although the pace of migrant arrivals to the Balkans had dropped since March 2016, hundreds of people continued to populate the area around the train and bus stations of the capital, finding shelter in occupied warehouses with no access to any facility, until they were evicted en masse in May 2017.[38]

Explaining the solidarity movement along the Western Balkans route: the emotional involvement

Following the cultural turn in the social sciences, which disclosed the prominent role that emotions play in political conflicts,[39] along with cognitive processes, organizational and structural factors, in this section I explore the emotional dynamics that informed the decision of individuals to engage in solidarity initiatives with migrants at the outskirts of the EU. I contend that the emotional involvement of activists and volunteers with the refugees compensated at least in part for the absence of networks and previous experience of volunteerism of individuals engaged in pro-refugee activism. As the following excerpts from my interviews show, a significant number of interviewees described their decision to engage in solidarity initiatives as grounded in humanity, outrage and compassion, which Jasper defined as 'moral emotions'.[40] According to Jasper, moral emotions consist of

> feelings of approval and disapproval based on moral intuitions and principles, as well as the satisfactions we feel when we do the right (or wrong) thing, but also when we feel the right (or wrong) thing, such as compassion for the unfortunate or indignation over injustice.[41]

As Goodwin and other scholars have noted, moral emotions intertwine with cultural meanings and cognitive understandings, which shape our knowledge of 'what is right or wrong, good or bad'.[42] Far from being irrational, emotions are said to be 'created and reinforced in narrative and discourses'[43] and therefore 'firmly rooted in moral and cognitive beliefs that are relatively stable and predictable'.[44] In line with the emphasis the cultural turn puts on the importance of emotions, my analysis stresses that a strong emotional commitment drove activists and volunteers' behaviour, inclining them to intervene in support of migrants. In turn, biographical experiences of displacement and uprooting proved important prerequisites for pro-refugee engagement, as they constituted preconditions for moral emotions to develop. In that regard, an interviewee living in Macedonia motivated her engagement as grounded on compassion and humanity, and her decision to get involved in the solidarity movement as triggered by an episode that, disrupting her daily routine, urged her into action. In her interview, she recounts:

> Suddenly, I found myself driving at night on the road from Skopje to Stip, when I saw them [the migrants] on the road. People were walking with children near me and I was

> shocked to see little children walking. When you witness such a scene it totally changes you. You cannot just keep going with the car. I stopped and asked them what they needed [...] and this is how the organization started.[45]

For other activists, the motivation to engage in the solidarity movement got reinforced during the action, thanks to the establishment of affective and emotional bonds developed by contact with migrants. The following excerpt of an interview with a lawyer working for the Macedonian Young Lawyers Association (MYLA) association in Skopje, a national organization active for about a decade in mobilization for refugees' rights—and in particular for Kosovars and Roma citizens, reports:

> I was working both in the office and in the camps since June. It has been very challenging to see so many people and to hear so many stories of war and traumatic experiences. What personally motivated me are the will, the motivation and the positivity the refugees had. They had been through so much and stay positive sometimes, and speak about such horrendous events. This motivated me to keep doing what I am doing. People are fleeing from horrendous situations and trying to have a bit of safety for themselves.[46]

Compassion, empathy and humanity likewise informed the personal decision to get involved in solidarity actions of a young psychologist who worked in the state registration camp of Preševo, in southern Serbia. She describes her commitment as emotionally grounded, but at the same time also emotionally charged:

> When somebody tells you about his life you understand him. You become close to him, and everything happening to him affects you. Now I see it as more personal than I should. Anything that happens now in Afghanistan and Pakistan affects me more than before. I got close to them [the migrants] and this had a lot of impact. Hearing that people die is not easy and it is always so distant, until you meet their culture. [...] I have different feelings: anger towards the EU and other organizations; sometimes I felt helpless and hopeless, we [practitioners and volunteers] had debriefing with psychologists in order to remain sane.[47]

Often, the emotional commitment stemmed from a process of identification of the individuals involved in solidarity actions with the group of migrants in transit, originating from the perception of having undergone a similar experience of displacement and/or uprooting in the past. In several cases, feelings of empathy and compassion originated from personal life experiences, which constituted another element partly accounting for altruistic commitment. As a matter of fact, a significant number of activists interviewed underwent forced displacement during the Yugoslav wars of the 1990s, and mentioned it as a factor motivating their engagement. An activist in Macedonia connected her decision to get involved in the solidarity movement with her personal experience of displacement in the recent past. As she points out, several people living in the region share this condition, as, she claims, they still remember clearly what it means to seek refuge from a war:

> I am a Bosniak, we had a refugee crisis ... I felt how war affects people. Also the Albanians [living in Macedonia] remember the refugee crisis in Kosovo, and the refugees used the same path the Macedonians used during the Great War when the Greeks expelled them.[48]

While, on the one hand, a personal background of displacement inclined individuals to take action, on the other a biographical experience of living abroad also informed the decision to get involved in collective solidarity actions in support of refugees. The following interviewee referred to her story of uprooting as a predictor of her commitment:

> I travelled a lot all my life, and lived in different countries for 11 years, so I know what it means to be displaced. When I saw the refugees crossing our country it was certainly my life experience that brought me to help them.[49]

Another NGO practitioner from Serbia portrayed her commitment as a moral duty in light of her personal biography, as her words highlight:

> As a foreigner, I was not forced to go outside my country [...] I decided to explore everything and I thought this was an occasion to discover people more. [...] I think the involvement of foreigners motivated the Serbs to get involved. For instance, in Miksalište some volunteers were young people doing the Balkan tour, who saw the refugees and ended up there.[50]

While appealing mostly to humanitarian principles as drivers of their commitment, the empirical analysis reveals that emotions had a prominent role, and often lay at the heart of the decision to get engaged in solidarity actions with migrants. Hence, a similar emotional involvement drove into action and brought together individuals some of whom were new to this kind of experience. The solidarity movement in both Serbia and Macedonia tapped into emotional resources to compensate for the lack, or weakness, of pre-existing solidarity networks and social ties, and the sparse organizational and financial capacities available to the movement. In this respect, emotional resources allowed the solidarity movement to prosper, compensating in part for the limited financial and organizational capacities of the movement.

The framing of the crisis: a discursive opening

The second factor that partially accounts for the flourishing of a refugee solidarity movement along the Western Balkans route concerns the discursive strategies employed by institutional actors during the period in which the Western Balkans corridor remained open. The collective narrative that officials put forward created fertile ground for the movement to develop, increasing the prospects for altruistic mobilizations. To that end, I explore the context in which the solidarity movement emerged and unfolded. In fact, social movements scholarship has resorted to the role that external factors play in the movements themselves to explain the occurrence of collective action, putting forward the concept of 'opportunities'.[51] Specifically, some scholars have explored the extent to which the economic and political context in which collective action occurs influences mobilization, creating opportunities for action.[52] Other scholars have expanded the rubric of 'opportunities' to the cultural sector, elaborating the notion of 'discursive opportunities'.[53] Essentially, this concept takes into account the cultural dynamics in which mobilization occurs, and the way in which a certain issue is framed within a certain polity at a specific time.[54] With this notion, scholars have aimed at exploring the cultural elements that, in a broader environment, facilitate/constrain successful social movement framing. By drawing on the concept of 'discursive opportunity structure', I thus analyse the ways in which government officials framed the refugee crisis during 2015's long summer of migration, and the extent to which this created favourable prospects for solidarity actors.

Throughout summer 2015, therefore, at the beginning of the 'crisis', state officials in Serbia and Macedonia referred to the growth in migration flows as a transient and exceptional phenomenon, in a similar fashion to Western European leaders.[55] In so doing, they enhanced and promoted an understanding of the migratory movement along the Western Balkans route mostly as a temporary and out-of-the-ordinary event. In their public discourses, domestic state authorities often underlined that the migrants did not constitute a severe threat to the local population, as they were not intending to settle in their territories. Hence, these narratives contributed to reinforce the perception of the Western Balkans as representing only a provisional stage along the route towards other EU countries, more attractive for migrants than the non-EU and economically poor former Yugoslav states. In that regard, Macedonia's President Gjorge Ivanov, of the conservative-nationalist party Internal Macedonian Revolutionary Organization – Democratic Party for Macedonian National Unity (VMRO-DPMNE), claimed publicly: 'The refugees are waiting for Merkel to pick them up in planes or trains. They all want to go to Germany and do not want to stay in Greece.'[56]

In a similar fashion, the then Serbian premier Aleksandar Vučić (Serbian Progressive Party, SNS) held public discourses about migrants, pointing to the exceptionality of the situation and emphasizing the country's willingness to provide medical assistance and other types of support necessary 'for the few days the migrants usually spend in Serbia'.[57] Similarly, the Serbian PM often openly praised the humanitarian governmental commitment towards migrants, supported on the ground by the well-disposed behaviour of the Serbian police, which showed tolerance towards the presence of refugees in the parks of Belgrade throughout the summer. The picture of a Serbian policeman, later disclosed to have Albanian origins, cuddling a Syrian child[58] went viral throughout social networks in autumn 2015, contributing to strengthen the image of tolerant domestic police forces in the media.[59] Besides claiming publicly the welcoming stance of Serbia towards migrants and refugees, in August 2015 the premier personally visited the parks of Belgrade, where migrants had found temporary respite. To reinforce Serbia's welcoming stance, the PM resorted to his refugee background, and to the country's welcoming policies towards IDPs in the aftermath of the 1990s Yugoslav wars.[60] On another occasion, he stressed the intention to leave the country's borders open by refusing to compare asylum seekers with terrorists, saying:

> We speak about desperate people, we don't speak about criminals and terrorists. They are just on their way to find better lives for themselves and their kids: they need help, they don't need condemnation and punishment. You cannot stop the flow of life with fences.[61]

In line with their narratives revolving around migration as a temporary and exceptional phenomenon, national governments of both Serbia and Macedonia favoured migrant transit with concrete policies. Both countries provided specially commissioned trains to facilitate the passage towards other destinations in the EU, or towards the next country's border. Furthermore, in 2015 the Serbian and Macedonian governments provided people on the move with 'semi-regular legal statuses',[62] introducing the so-called 'white cards', documents allowing migrants to legally cross the former Yugoslav countries within a 72-hour time frame. Moreover, both governments publicly declared they had no possibility of hosting a high number of potential asylum seekers in their territories, as the temporary hosting facilities in Macedonia were not suited to accommodate people for an extended stay, having a capacity of only 2000.[63] In a similar fashion, in Serbia the centres aimed at hosting asylum seekers near Belgrade still had to be emptied of the IDPs who had been living there since the end of the 1990s war. The official statistics seemed to mirror the transitory nature of the passage through the Western Balkan countries, as the numbers of individuals lodging an asylum application in Serbia or Macedonia in 2015 and 2016 were insignificant compared with Austria, a destination country. As at 31 December 2015, 557,795 refugees and migrants expressed their intention to seek asylum in Serbia,[64] while 43,590 submitted an asylum application in Macedonia.[65] These data might seem misleading though, as both former Yugoslav countries required migrants to register their intention to seek asylum in the country in order to cross their territories. However, the majority of them abandoned the country before effectively lodging a claim. For instance, data show that, in 2015, only 30 positive decisions were taken in response to the 557,995 asylum applications submitted in Serbia, and three in Macedonia out of 43,590 asylum claims received in the same year.[66] By contrast, in Austria, during the same period, the number of people claiming international protection amounted to 88,160, of whom 21,045 received a positive answer.[67]

Such a 'temporary humanitarian approach'[68] of the former Yugoslav states towards refugees provided a fertile ground for humanitarian grassroots initiatives to prosper, favouring the emergence of groups providing first-hand help to the migrants, at times in cooperation with local authorities. In several instances, it emerged that the discourses of solidarity actors aligned with the interpretation of the domestic authorities, portraying the migratory movement along the Western Balkans route as a transient phenomenon. In that regard, one interviewee associated the tolerance and welcoming stance of locals towards asylum seekers owing to the alleged temporality of the migratory phenomenon, by saying: 'People were tolerant as they [migrants] did not want to stay there'.[69] In a similar fashion, another interviewee claimed that: 'Serbia sees itself as a transit country [...] People are nice because they know migrants would leave the country'.[70] The fact that the majority of solidarity actors framed themselves as fulfilling a humanitarian-oriented rather than political-oriented type of action, owing also to their inability to elaborate a political strategy due to the emergency situation they were facing, contributed to the perception of solidarity initiatives as not representing a threat to local authorities, at least in the period preceding the corridor's shutdown. This attitude, in turn, facilitated the collaboration with state authorities, which relied on solidarity actors to handle the influx of refugees—or, better, their transit to the next country—especially in the first stage of the refugee crisis.

In spite of taking slightly different approaches to the refugee crisis, Serbia and Macedonia both depicted themselves as protecting the EU's borders, notwithstanding them not being EU member states and claiming not to receive enough financial support from the EU. Another discursive strategy adopted in 2015 by institutional actors in the two former Yugoslav states consisted in portraying the countries as fulfilling the mission of acting as bulwarks against the security threat migrants were allegedly posing to Europe. In that regard, the President of Macedonia Ivanov stated publicly that: 'The crisis is coming from the territory of Greece—a Schengen area state. So Macedonia has to defend Europe from the EU itself.'[71]

Government officials in both countries described their states as victims of the EU system, paying for the inability of the EU to cope with migrant arrivals, while receiving little in return. Although representing migrants mostly as a security concern (as was the case with Macedonian officials), the narrative in both countries focused on the attribution of blame to the EU, incapable of coping with the rising number of refugees. Conversely, both countries' government officials praised the humanitarian response of local authorities and the altruistic behaviour of their citizens. Repeatedly, Macedonian President Gjorge Ivanov blamed the EU for not providing enough financial resources for the country to handle the arrivals, while giving substantial financial help to Greece, a member of the EU, which proved, in his view, unable to put a halt to migrant arrivals in the neighbour country. In one occasion, he said: 'As a transit-only country, Macedonia is doing this for the benefit of the EU. Unfortunately, even as we are trying to implement EU decisions, we face unfair criticism for doing so.'[72]

In other speeches, President Ivanov stressed the 'sacrifice' of Macedonia, 'paying for the mistakes of the EU'.[73] He also pointed to Greece as the true culprit regarding the transit of refugees, and praised the humanitarian stance of its citizens, although stating clearly that Macedonia would not pay for the EU's mistakes. In that regard, he said:

> We are also showing a humane face. The Macedonians have provided an incredible amount of help. They have provided medical care and made donations. But once again: We are not a EU country and have to pay for the EU's mistakes.[74]

Similarly, the Serbian PM prided himself on the compassionate and humane treatment of refugees crossing Serbian territory, claiming to be 'satisfied and proud' of the humanitarian behaviour of ordinary citizens.[75] The Serbian welcome policy is best understood in relation to the country's prospects of EU accession.[76] In Serbia, the initial pro-refugee stance of the government was coupled with a commitment to comply with EU standards, in order to accelerate Serbia's path towards EU membership. In an attempt to demonstrate goodwill and commitment to European values, premier Vučić pointed out the commitment of the country in supporting refugees by claiming that the country was 'more European that the Europeans themselves'.[77] In so saying, the PM wished to improve the country's image internationally,[78] underlining its willingness and commitment to follow EU rules, notwithstanding the financial constraints derived from not yet being part of the EU, in an attempt to increase its leverage over the EU. Essentially, the narrative coupling the blame directed at the EU and certain EU countries, claimed out as responsible for the mismanagement of migrant arrivals, with the praise for local citizens and state authorities in handling the humanitarian crisis during summer 2015, as well as the alignment of the solidarity actors' discourse with the authorities' narratives, provided a favourable context for a solidarity movement and its self-organized relief efforts to flourish.

Conclusion

The opening of the Western Balkans corridor represented the first time after the end of the 1990s' wars that the former Yugoslav states found themselves handling a massive arrival of people, which seemingly caught policy-makers unprepared to handle it. By contrast, the migrant flow did not catch the volunteers, activists and grassroots solidarity groups by surprise. Notwithstanding the chaotic and rapidly changing situation at the border crossings, solidarity actors provided timely humanitarian assistance to support transit migrants, at times filling a void created by the absence of an appropriate response by state authorities. Not only did pre-existing groups mobilize, while extant ones shifted their focus to meet the refugees' immediate needs, but new ones also emerged and spread, while new alliances and collaborative networks developed at both the national and transnational level.

Essentially, this article has shed light on two factors in particular, which accounted for the emergence of local solidarity initiatives in support of migrants crossing the Western Balkans in summer 2015. First, the study revealed that emotional resources played an important role, as moral and emotional questions informed the decision of individuals to get involved in migration-related political activism, while biographical experiences of displacement or uprooting represented a predictor of solidarity actors' commitment. Secondly, it showed that the evolution of solidarity movements was strongly shaped by discursive opportunities, particularly at the beginning of the refugee crisis. At first, in fact, governments in Macedonia and Serbia portrayed the inflow of migrants as a temporary rather than a long-term phenomenon, owing, in their opinion, to the peculiar position that the Yugoslav successor states occupy in the geography of migration, and them being less attractive for migrants vis-à-vis EU countries. Thus, local governments framed the passage of migrants by adopting a logic of exceptionality and temporality, which contributed to create a feeling of empathy towards migrants, as they favoured the perception of migrants as not posing a severe threat to the local population, for they did not intend to settle in the former Yugoslav countries. The humanitarian approach of local authorities, accompanied by policies favouring the transit rather than the reception of asylum seekers, increased the prospects for altruistic mobilizations in both countries at the gate of the EU.

The scenery changed dramatically with the progressive closing of borders and the shutdown of the corridor following the EU–Turkish agreement in March 2016. As a consequence, the flow of refugees lost intensity progressively throughout the first months of 2016, and the number of arrivals in the Western Balkans fell sharply. The situation became particularly tense in Belgrade, where hundreds of refugees got stuck, unable to continue their path forward, owing to their inability to cross both the Croatian and the Hungarian borders. The narrative twisted accordingly. Humanitarianism faded from public discourses, and the activities of NGOs and solidarity groups began to be further obstructed by different means. After the shutdown of the corridor, both governments hampered the self-organized provision of assistance to stranded migrants. In Belgrade, local solidarity groups were prevented from distributing meals to the migrants in the parks in November 2016. The situation worsened to the extent that one interviewee claimed: 'last year [2015] they had tents and toilets, now they have nothing'.[79] On 9 March 2016, Macedonia closed its border with Greece, while the Serbian PM repeated on many occasions that 'Serbia won't serve as a parking lot for refugees that the EU does not accept'.[80] The attitude towards the migrants also deteriorated in the media. Notwithstanding the changed scenario and the progressive securitization of the borders, grassroots solidarity initiatives continued along the Western Balkans route by different means even after the official shutdown of the corridor.

Notes

1. UNHCR, 'Internally displaced people' <http://www.unhcr.org/internally-displaced-people.html> (accessed 6 March 2018).
2. Norwegian Refugee Council—Internal Displacement Monitoring Centre, 'Bosnia and Herzegovina. Ethno-political agendas still prolonging displacement', 2014 <https://reliefweb.int/sites/reliefweb.int/files/resources/201411-eu-bosnia-overview-en.pdf≥ (accessed 18 November 2018).
3. Internal Displacement Monitoring Centre, 'Serbia', 2015 <http://www.internal-displacement.org/countries/serbia≥ (accessed 13 November 2018).
4. European Commission, 'Managing the refugee crisis. Western Balkans route: state of play report', 10 February 2016 <http://ec.europa.eu/dgs/home-affairs/what-we-do/policies/european-agenda-migration/background-information/docs/western_balkans_route_state_of_play_report_en.pdf> (accessed 5 July 2017).
5. While mass-media reports usually adopted the term 'refugees' to refer both to asylum seekers and migrants, I acknowledge that the terms 'refugee' and 'asylum seeker' have different meanings. For the sake of brevity, throughout the article I use the expression 'migrant', 'refugee', 'asylum seeker' and 'people on the move' interchangeably to refer to individuals who have fled their countries in a bid to escape war, or as a result of economic deprivation, regardless of having lodged an asylum claim or having been granted international protection under the 1951 Refugee Convention on the Status of Refugees UNHCR, 'Text of the 1951 Convention Relating to the Status of Refugees', 1951 <http://www.unhcr.org/3b66c2aa10> (accessed 8 March 2018).
6. Balkans in Europe Policy Advisory Group, 'Policy brief: The migrant crisis: a catalyst for EU enlargement?', *Www.Biepag.Eu*, June 2016 <http://www.balkanfund.org/publib/biepag/The-migrant-crisis-a-catalyst-for-EU-enlargement-web.pdf> (accessed 13 November 2018).
7. Ibid.
8. Senada Šelo Šabić and Sonja Borić, 'At the gate of Europe. A report on refugees on the Western Balkan route', March 2016, Friedrich Ebert Foundation edition <http://library.fes.de/pdf-files/bueros/kroatien/13059.pdf> (accessed 13 November 2018).
9. Vedran Dzihic, 'Border (power) politics reborn in the refugee crisis: contested lines of demarcation and the re-emergence of identity conflicts: the case of Serbia and Croatia', in *Border Politics*, Springer, Cham, 2017, pp. 51–63 <https://doi.org/10.1007/978-3-319-46855-6_4>.
10. Aleksandar Pavlović, 'A passage to Europe: Serbia and the refugee crisis event analysis', *Contemporary Southeastern Europe* 3(1), 2016, pp. 59–65.
11. Throughout the article I refer to the country interchangeably as Macedonia or FYROM, acronym for 'Former Yugoslav Republic of Macedonia', the provisional name of the country in use by international organizations and states until the solution of the controversy with Greece regarding the naming of the country.
12. BBC news, 'Macedonia migrants: hundreds rush border', 21 August 2015 <http://www.bbc.com/news/world-europe-34014353> (accessed 8 March 2018).
13. President of Serbia since May 2017.
14. As the transit of migrants was organized and conducted under the supervision of domestic authorities and international organizations.
15. The Schengen Borders Code provides member states with the ability to temporarily reintroduce border controls at internal borders in the event that a serious threat to public policy or internal security has been established (European Commission, Migration and Home Affairs).
16. Donatella della Porta (ed.), *Solidarity Mobilizations in the 'Refugee Crisis': Contentious Moves*, Palgrave, Basingstoke, 2018.
17. Interview with a member of Refugee Aid Serbia, Belgrade, 28 July 2016.

18. I refer to solidarity movements as those movements that do not benefit directly from the outcome of their involvement, whose actions can be defined as altruistic and political at the same time (Marco Giugni and Florence Passy (eds), *Political Altruism? Solidarity Movements in International Perspective*, Rowman and Littlefield, Lanham, MD, 2001).
19. The interviews were conducted in the framework of the ERC-funded project 'Mobilizing for Democracy: Collective action and the refugee crisis' <http://cosmos.sns.it/collective-action-and-the-refugee-crisis/> (accessed 10 October 2018).
20. Given the sensitivity of the topics and owing to the potentially easy identification of the informants, I have chosen not to report interviewees' private data.
21. Kiri Sandter and Vera Wriedt, '(De-)constructing borders contestations in and around the Balkan corridor in 2015/16', *Movements. Journal für Kritische Migrations- und Grenzregimeforschung*, 3(1), 2017, pp. 141–150.
22. As noted above (see note 15), The Schengen Borders Code provides member states with the ability to temporarily reimpose border controls at internal borders in exceptional circumstances.
23. European Commission, 'Commission Staff Working Document: The Former Yugoslav Republic of Macedonia Report 2015. Accompanying the Document Communication from the Commission to the European Parliament, the Council, the European Economic and Social Committee and the Committee of the Regions', 2015, <https://ec.europa.eu/neighbourhood-enlargement/sites/near/files/pdf/key_documents/2015/20151110_report_the_former_yugoslav_republic_of_macedonia.pdf> (accessed 10 March 2018).
24. Ibid.
25. Ibid.
26. Associated Press in Skopje, '14 Migrants killed by train while walking on tracks in Macedonia—police', *The Guardian*, 24 April 2015 <https://www.theguardian.com/world/2015/apr/24/several-migrants-hit-by-train-killed-central-macedonia> (accessed 9 March 2018).
27. Interview with an administrator of Help the Refugees Macedonia FB page, 13 June 2016, via Viber; Interview with a representative of NGO Legis, 20 July 2016, Skopje.
28. Interview with an administrator of Help the Refugees Macedonia FB page, 13 June 2016, via Viber.
29. Interview with a spokesperson of MYLA, 21 July 2016, Skopje; Interview with a representative of the NGO Legis, 20 July 2016, Skopje.
30. The refugees had to pay for their own transportation to the Serbian border, and the price of the railway ticket was increased from €7 to €25, allowing the Macedonian railway company to make millions in profits.
31. Irina Chudoska Blazhevska and Carlos Flores Juberias, 'Macedonia in the 2015 Refugee Crisis', *Balkania*, 7, 2016, pp. 217–244.
32. Šelo Šabić and Borić, op. cit.
33. The two main parks of the capital were renamed 'Afghani' and 'Syrian' parks, owing to the presence of migrants from those two countries that gathered in the green areas according to their nationalities.
34. European Civic Forum, 'European Democratic Citizenship Award. Refugee Aid Miksalište Campaign', <http://civic-forum.eu/wp-content/uploads/2016/06/Refugee-Aid-Miksaliste-Campaign.pdf> (accessed 13 November 2018).
35. Open Borders, 'Report from Belgrade', 1 April 2016 <https://openborder.noblogs.org/post/2016/04/12/report-from-belgrade/> (accessed 10 March 2018) Lazara Marinković, 'Poslednji Dan u Skvotu Za Izbeglice u Beogradu [The last day in the squat with refugees in Belgrade] | VICE | Srbija', VICE, 3 May 2016 <http://www.vice.com/rs/read/poslednji-dan-u-noborder-hostelu-skvotu-za-izbeglice-u-beogradu> (accessed 10 March 2018).
36. No border Serbia, 'Zidovi, Zidovi … ograde, Ograde…', [Walls, walls … fences, fences …] No border Serbia, 8 December 2016 <https://noborderserbia.wordpress.com/> (accessed 10 March 2018).

37. The Commissariat was originally created in the 1990s to handle internally displaced persons (IDPs) from the Yugoslav wars. Since 2012 it is also in charge of managing migration <http://www.kirs.gov.rs/articles/aboutus.php?lang=ENG> (accessed 11 March 2018).
38. Jelena Bjelica and Martyne van Bijlert, 'Afghan exodus: the re-emergence of smugglers along the Balkan route—Part III', *Afghanistan Analysis Network* (blog), 10 August 2016 <https://www.afghanistan-analysts.org/afghan-exodus-the-opening-and-closing-of-the-balkan-corridor/> (accessed 6 March 2018); 'Refugees in Serbia—stranded in a warehouse', *Deutsche Welle*, 17 January 2017 <http://www.dw.com/en/refugees-in-serbia-stranded-in-a-warehouse/a-37167779> (accessed 6 March 2018).
39. See, amongst others, William A. Gamson, *Talking Politics*, Cambridge University Press, Cambridge, 1992; James M. Jasper, 'The emotions of protest: affective and reactive emotions in and around social movements', *Sociological Forum*, 13, 1998, pp. 397–424; Helena Flam, 'Emotions map: a research agenda', in Helena Flam and Debra King (eds), *Emotions and Social Movements*, Routledge, London, 2005, pp. 19–41.
40. James M. Jasper, 'Emotions and social movements: twenty years of theory and research', *Annual Review of Sociology*, 37, 2011, pp. 285–303.
41. Ibid.
42. Jeff Goodwin, James M. Jasper and Francesca Polletta, 'Emotional dimensions of social movements', in David A. Snow, James M. Jasper and Francesca Polletta (eds), *The Blackwell Companion to Social Movements*, Blackwell Publishing, London, 2004, p. 422.
43. Gamson, op. cit., p. 423.
44. Jasper, op. cit., p. 421.
45. Interview with the administrator of the Help the Refugees in Macedonia FB platform, via Viber, 13 June 2016.
46. Interview with the spokesperson of Macedonian Young Lawyers Association, Skopje, 21 July 2016.
47. Interview with a volunteer of ADRA, Belgrade, 25 July 2016.
48. Interview with the spokesperson of ONG LEGIS, Skopje, 20 July 2016.
49. Interview with the administrator of the Help the Refugees in Macedonia FB platform, via Viber, 13 June 2016.
50. Interview with a cultural mediator working for NGO ADRA in the state-run camp of Preševo, Belgrade, 25 July 2016.
51. H. Kitschelt, 'Political opportunity structures and political protest: anti-nuclear movements in four democracies', *British Journal of Political Science*, 16(1), 1986, pp. 57–85; Hanspeter Kriesi Ruud Koopmans, Jan Willem Dyvendak, and Marco Giugni (eds), *New Social Movements in Western Europe: A Comparative Analysis*, University of Minnesota Press, Minneapolis, 1995; Sidney Tarrow, *Power in Movement: Social Movements, Collective Action and Politics*, Cambridge University Press, Cambridge, 1998.
52. Tarrow, op. cit.
53. Paul Statham and R. Koopmans, 'Ethnic and civic conception of nationhood and the differential success of the extreme right in Germany and Italy', in *How Social Movements Matter*, University of Minnesota Press, Minneapolis, 1999; Ruud Koopmans and Susan Olzak, 'Discursive opportunities and the evolution of right-wing violence in Germany', *American Journal of Sociology*, 110(1), 2004, pp. 198–230.
54. Statham and Koopmans, op. cit.
55. Pietro Castelli Gattinara, 'The "Refugee crisis" in Italy as a crisis of legitimacy', *Contemporary Italian Politics*, 9(3), 2017, pp. 318–331.
56. *Business Insider*, 'Macedonian president to Germany: Your country has completely failed', 11 March 2016 <http://www.businessinsider.com/macedonian-president-to-germany-your-country-has-completely-failed-2016-3?IR=T≥ (accessed 3 February 2018).
57. Blic.rs, 'Vučić Migrantima: Ovde Ste Uvek Dobrodošli, Osećajte Se Sigurno' [Vučić to migrants: here you are always welcome, feel safe], 19 August 2015 <http://www.blic.rs/

vesti/drustvo/vucic-migrantima-ovde-ste-uvek-dobrodosli-osecajte-se-sigurno/ncwns15> (accessed 15 March 2018).

58. BBC, 'Why this picture of a migrant child being hugged went viral', *BBC News*, 10 September 2015, sec. Trending <http://www.bbc.com/news/blogs-trending-34210994> (accessed 11 March 2018).
59. Chiara Milan and Andrea L. P. Pirro, 'Interwoven destinies in the "long migration summer": solidarity movements along the Western Balkan route', in Donatella Della Porta (ed.), *Solidarity Mobilizations in the 'Refugee Crisis'*, Palgrave Macmillan, Cham, 2018, pp. 125–153.
60. Al Jazeera Balkans, 'Vučić: Izbjeglice Su Ljudi, a Ne Vanzemaljci' [Vučić: refugees are human beings, not aliens], 19 September 2015 <http://balkans.aljazeera.net/vijesti/vucic-izbjeglice-su-ljudi-ne-vanzemaljci> (accessed 8 March 2018).
61. Reuters, 'Serbia won't close borders to migrants', 26 August 2015 <https://af.reuters.com/article/worldNews/idAFKCN0QV2AM20150826≥ (accessed 19 February 2018).
62. Julija Sardelić, 'From temporary protection to transit migration: responses to refugee crises along the Western Balkan route', *EUI Working Papers* RSCAS 2017/35 (2017).
63. Chudoska Blazhevska and Juberias, op. cit.
64. UNHCR and UN Serbia, *Serbia, Interagency Operational Update*, December 2015 <http://www.statewatch.org/news/2016/jan/unhcr-serbia-dec-jan.pdf> (accessed 14 May 2018).
65. European Commission, 'Commission staff working document: The Former Yugoslav Republic of Macedonia Report 2015. Accompanying the document Communication from the Commission to the European Parliament, the Council, the European Economic and Social Committee and the Committee of the Regions', 2015 <https://ec.europa.eu/neighbourhood-enlargement/sites/near/files/pdf/key_documents/2015/20151110_report_the_former_yugoslav_republic_of_macedonia.pdf> (accessed 10 March 2018).
66. Bodo Weber, *The EU–Turkey Refugee Deal and the Not Quite Closed Balkan Route*, Friedrich-Ebert-Stiftung in Southeast Europe, Sarajevo, 2017 <http://library.fes.de/pdf-files/bueros/sarajevo/13436.pdf> (accessed 15 February 2018).
67. European Migration Network, 'Country factsheet: Austria 2015', 2015 <http://www.emn.at/wp-content/uploads/2016/12/Country-Factsheet-Austria-3.pdf≥ (accessed 15 May 2018).
68. Sardelić, op. cit.
69. Interview with a volunteer of ADRA, Belgrade, 25 July 2016.
70. Interview with a journalist of Radio Free Europe (*Radio Slobodna Europa*), Belgrade, 25 July 2016.
71. Gjorge Ivanov, 'Macedonia is defending Europe from itself', 6 March 2016, sec. World <https://www.telegraph.co.uk/news/worldnews/europe/macedonia/12185464/Macedonia-is-defending-Europe-from-itself.html> (accessed 15 May 2018).
72. Ibid.
73. Charlotte Beale, 'Refugee crisis: Macedonia tells Germany they've completely failed', *The Independent*, 12 March 2018 <https://www.independent.co.uk/news/world/europe/macedonia-tells-germany-youve-completely-failed-a6927576.html> (accessed 15 May 2018).
74. Ibid.
75. Al Jazeera Balkans, op. cit.
76. Milan and Pirro, op. cit.
77. Pavlović, op. cit.
78. Ibid.
79. Interview with a journalist of Radio Free Europe (*Radio Slobodna Europa*), Belgrade, 25 July 2016.
80. B92.net, 'PM: Serbia won't be parking lot for migrants—on B92.Net', B92.net, 3 February 2016 <http://www.b92.net/eng/news/politics.php?yyyy=2016&mm=02&dd=03&nav_id=96915> (accessed 12 March 2016).

Disclosure statement

No potential conflict of interest was reported by the author.

Funding

This work was based at the Scuola Normale Superiore in Florence and supported by the European Research Council (ERC) [grant number 269136] 'Mobilizing for Democracy: Collective action and the refugee crisis' (project description available at: http://cosmos.sns.it/collective-action-and-the-refugee-crisis).

ORCID

Chiara Milan http://orcid.org/0000-0002-2604-3442

The Floating Pumpkin Syndrome: Forced Migration, Humanitarian Aid, and the Culture of Learned Helplessness

Nina Bosankić, Enisa Mešić and Bojan Šošić

ABSTRACT
Since the 1992–1995 war, Bosnia and Herzegovina has been going through a difficult process of transition, characterized by dependence on humanitarian aid and by major challenges regarding employment. Returnees and displaced persons have experienced many unfavourable circumstances and challenges. This paper draws from the findings gathered by the authors over several years of field research with more than 2000 returnees and internally displaced persons (IDPs) in Bosnia and Herzegovina, mostly in areas with high return rates, including social housing beneficiaries. It offers a reflection of contextual factors in shaping the world of migrants and their potential to cope and adapt, and the role of donor agencies in imposing 'learned dependency'. Using the metaphor of a pumpkin that floats on the water, following its currents, but retaining own integrity, the paper emphasizes that it may actually be the impact of adversity that awakens certain qualities and skills in people. Processes analogous to post-traumatic growth may be in effect in increasing the potential of the individual to develop a silent and potential form of understanding the world around oneself, a sort of 'personal wisdom'. Hence, a positive psychology approach to studying the personal effects of forced migration wisdom is advocated.

'we must first and foremost rid ourselves of the very Western, bourgeois, and hence very disparaging, idea that the masses are incapable of governing themselves'.

(Frantz Fanon, 1961/2004)[34]

Introduction

There is a traditional curse in Bosnia, probably formulated on grounds of the personal experience of the hardship it imposes on the target of the curse: 'May you have, and then not' (in Bosnian: '*Dabogda imao pa nemao!*'). This, perhaps as a sort of a self-fulfilling prophecy, may well be the briefest summary of the effects of the war that struck the people of Bosnia and Herzegovina (BiH) at the end of the last century, between 1992 and 1995. The first post-war census (2013) showed enormous demographic changes,[1] and the migrations, now voluntary, continue, mostly to West European countries. The country has lost its demographic character that had earned its map the metaphor of leopard's skin for the high degree of diversity.[2]

An industrialized country on the brink of transition, with an already established stock market, sank into war, with tremendous demographic and economic consequences. For a majority of individuals and families, the changes in their quality of life may be said to reflect the change described in the curse cited above. The war itself, as well as the attendant forced migration within or out of the country, led to substantial uprooting[3] and loss of the so-called 'assumptive world', that is, fundamental assumptions about the world and reality of past and future events based on previous experiences.[4] A broader manifestation of such personal realization of change comes regionally in the form of Yugo-nostalgia: the longing for the economic security, sense of solidarity, socialist ideology, multiculturalism, internationalism and non-alignment, history, customs and traditions, and a better way of life; as a 'cultural product of a specific economic context of post-socialist societies that has spread across the Balkans'.[5]

The effects were different on various scales. On the group scale, public discourse, especially by prominent political figures, was in line with the demographic changes resulting from the war and the Dayton Accord, often acquiescing to the current demographic map and quietly opposing the agenda for return. 'With breaking the secular ties in the post-Soviet and the post-colonial era, local leaders have sought new grounds to build social identities and political loyalty and many turned to ethnicity and religion.'[6] It is worth noting that these '[i]nter-religious tensions did not start until the late nineteenth century and were incited from outside the country: namely by the Greater-Serbian and Greater-Croatian religious-national mythologies and nationalistic programmes'.[7] Nevertheless, these effects seem not only to stick, but to have a tendency of aggravation, thanks to the formulations, or at least interpretations of the agreement that stopped the war: 'While thick conceptions of national identity and a lack of tolerance of dual nationality dominate the contemporary Bosnian scene, the final legacy of the Dayton Accord may be something much more disturbing: ethnic segregation and the consolidation of ethnically homogenized entities and cantons.'[8] The situation in a psychosocial sense is far from clear, not providing much opportunity to offer exact predictions about the future, since the Bosnian-Herzegovinian society is 'still facing the continual fabrication of ethno-nationalism, the wide-spread denial of atrocities and impunity, initiating trauma transformation, the repair and restoration of relationships, rebuilding trust and strengthening reconciliation in a tense post-conflict arena is difficult, but hope-infusing and promising'.[9]

Since the end of the war that battered the country from 1992 until 1995, rigid ethnic divisions have become the dominant characteristic of the political and social spheres in Bosnia and Herzegovina. Namely, the discursive 'ancient hatreds' and 'irreconcilable' ethnic differences which were used to polarize discourses and incite conflict and the eventual war, turned into real individual tragedies and painful experiences during the war, and later resulted in real intolerance, dislike and distrust.[10] The consequences are present in all areas of everyday life. Bosnia is characterized by exclusive, rather than inclusive, definitions of 'national community'; 'thick', rather than 'thin', conceptions of 'national identity', 'loyalty to one's nation' and 'distrust of ethnic others'; a lack of tolerance of dual nationality; and an almost non-existent public space shared with members of different national groups (as the result of the ethnic homogenization of major Bosnian cities).[11] It is also worth noting that Bosnia and Herzegovina is not only politically complex, but is also an economically disunited space, regulated by a decentralized and still not well harmonized legal framework, acting on several levels of jurisdiction (the implementation of relevant laws in the Federation of Bosnia and Herzegovina rests on individual cantons). According to data from the population census in 1991, BiH had a population of 4.3 million citizens, but during the war 2.2 million people left their pre-war homes: about 1.2 million fled the country and 1 million were internally displaced.[12] It is estimated that at the end of 2016 there were still 98,574 people in BiH with the status of internally displaced person (IDP).[13] Regardless of the fact that the war ended two decades ago and the rights to return to their pre-war property and to property restitution are specified in Annex VII of the General Framework Agreement for Peace in Bosnia and Herzegovina, the problem of displaced persons is still not entirely resolved. Although 'the post-1995 development policies advocated an active role for returnees and refugee migrants in promoting economic development in BiH',[14] many returnees sold property immediately upon their return, and then moved back to the place of displacement, usually that part of Bosnia where they could more easily integrate into the majority, or even went abroad. After the war, most people in BiH felt they were worse off than they had been before and, perhaps even more important, worse off than they had ever imagined they would be.[15] In the post-war period, poverty has been one of the most pressing issues, and displaced persons (in particular the elderly) as well as returnees are among the most vulnerable populations when it comes to the risk of falling into poverty and stigmatization. Poverty in the resident population and high unemployment have also prevented solidarity on a long-term basis. In 2017 BiH was ranked as first in the world for youth unemployment (ages 15–24), with a rate of 62.8%.[16] The effects of enforced migration and the problems arising from living as a returnee will continue to need to be dealt with in the future. At the same time, except for several anthropological studies,[17] practically no mental health studies have specifically addressed the population of forced migrants,[18] apart from one that also involved participation of the authors of this paper.[19] Most studies were conducted solely from a clinical perspective.[20] Article 4 of Annex VII in *The General Framework Agreement for Peace in Bosnia and Herzegovina: Agreement on Refugees and Displaced Persons* from 1995 postulates that 'The Parties shall not interfere with the returnees' choice of destination, nor shall they compel them to remain in or move to situations of serious danger or insecurity, or to areas lacking in the basic infrastructure necessary to resume a normal life.' The Ministry of Human Rights and Refugees of BiH, in an attempt to better tackle the protracted displacement situation, introduced in 2010 a *Revised Strategy of Bosnia and Herzegovina for the Implementation of Annex VII of the Dayton Peace Agreement* with a new shift in focus—from the right to return to local integration (including social housing and nursing home options), which means that the authorities had grasped the fact that many people simply opt to stay in the new community, rather than to insist on returning by all means. Originally adopted in 1999[21] (entity laws on displaced persons and returnees were adopted in 1995, harmonized with the state law), a revised state law on displaced people and returnees that incorporated new provisions for property compensation[22] was sent to the Council of

Ministers for adoption, but has remained stuck in the parliamentary procedure since 2012.

On the other hand, the people on an individual/family scale developed and manifested scarcely studied skills of survival in this sort of adverse environment. The term commonly used for this set of skills needed for the person and their family to cope and physically survive in a situation with poor chances of regular income, is *snalaženje*, which could be translated as 'managing for oneself', 'finding one's way out', 'getting along' or 'getting by'. This often includes moonlighting, ignoring regulations or even laws, exchanging favours, and in general relying on one's social network. A similar phenomenon, also related to dishonest strategies and evading responsibility, has been described in relation to certain periods of life in predominantly poor and rural parts of Italy,[23] and may also refer to how Italian migrants originally used to get along in new settings.[24]

Humanitarian assistance

During the four years of war, the Office of the United Nations High Commissioner for Refugees (UNHCR) and numerous international non-governmental organizations, including religious bodies, provided humanitarian aid worth millions of dollars to citizens of BiH, to locals and internally displaced persons alike. The international community kept providing humanitarian aid in one form or another for years after the Dayton peace agreement was signed: 'the more intense period of international involvement, between 1996 and 2000, brought in funds for infrastructure, refugee return, economic and structural reforms, and peacekeeping operations that, taken together, accounted for at least $15 billion worth of assistance'.[25] It was literally impossible for the vast majority of the country's population to make ends meet without humanitarian assistance.

Individual and communal perception of the aid received changed dramatically from time to time according to the situation in terms of personal experiences/challenges and the availability of resources being aggravated or ameliorated. Thus, although dependency on receiving humanitarian assistance was often perceived as a humiliating and necessary evil, it was also at times considered a life-saving injection. A common coping strategy to deal with the adversity associated with such humiliation and frustration was to resort to verbal revenge in the form of black humour[26]—for example, army ration hardtack was called *keramičke pločice* (ceramic tile). Even now, more than 20 years since the war ended, many people still have very vivid memories of humanitarian aid accompanied by mixed feelings towards such assistance. As was already observed in a study on humanitarian assistance focusing on the experiences of the citizens of Sarajevo, the affective power of past experiences remains captured in positive and negative memories and emotions towards the objects provided in humanitarian aid.[27]

Once the war ended, most people could not wait to be back on their feet, provide for themselves and their families and not depend on the international community's good will. 'Having survived adversity and many struggles, refugees tend to be resilient and resourceful and, if they encounter reasonably facilitative conditions, the majority of them can and do manage on their own with minimal or no assistance.'[28] Yet years of relying solely on humanitarian assistance and the failure by the state and international donors to introduce a comprehensive exit strategy that would facilitate bridging the gap between humanitarian emergency relief programmes and long-term development programmes in order to support and sustain the empowerment of vulnerable groups, led in part to the creation of a 'culture of dependence'[29] and 'learned helplessness'.[30] These syndromes were particularly widespread among internally displaced persons, especially senior citizens living in the collective centres. It seems that, '[i]nstead of focusing their activities on improving skills or on education activities, which could result in access to incomes through employment, international donors were focused on delivering humanitarian aid which only produced dependency of their beneficiaries'.[31] The unfavourable legacy—an underdeveloped economy, ethno-nationalism, the traumatizing past, non-existent public policy or budget funds to fight poverty and social exclusion—further contributed to many people and groups that were dealing with protracted displacement falling into lethargic contentment. They opt not to leave their safe zones, even if offered an altogether better alternative that requires just a bit more effort on their side. As a respondent in one of our field studies suggested, 'living in a collective centre was better [than in social housing]—because there we had flour and heating that someone else paid for'.

Field experiences

Over more than a decade of field research conducted by the authors, mostly jointly and with an approach of collecting both quantitative and qualitative data with more than 2000 returnees and internally displaced persons in BiH, lead us to develop a number of insights pertaining to the issues surrounding how the migrants are helped, and what their challenges are. These studies were conducted mostly in areas with a high return rate, including social housing beneficiaries. They offer an opportunity to reflect on contextual factors in shaping the world of migrants and their potential to cope and adapt. The following data, collected particularly through three different projects investigating social inclusion of returnees and displaced persons living in different types of accommodation, provided us with greater depth of understanding of migrants' life trajectories in the face of great adversity. We are referring to quantitative and qualitative data obtained through (i) a face-to-face survey from a convenience sample of 566 households with 1909 returnees from three municipalities with high rates of return—Bijeljina, Kozarac and Sanski Most; (ii) a mail survey, focus groups and semi-structured interviews from a purposive sample of 128 households with around 300 social housing beneficiaries, both returnees and displaced persons, mostly collective centre residents, from Srebrenica, Prijedor, Drvar, Goražde, Jablanica, Mostar, Fojnica and Foča; and (iii) an ethnographic study with a group of displaced women from eastern Bosnia residing in Sarajevo.[32] The quotes presented throughout the article were selected for being highly illustrative in terms of recurrent themes in participants' narratives.

We have often encountered a particular bias in post-war BiH, where many people seem to have become disposed to focus only on the negative aspects of their livelihood and lack of opportunities. Psychologically speaking, this phenomenon is often approached in terms of 'learned helplessness', which may be defined as a perceived absence of control over the outcome of a situation. In other words, people learn to behave helplessly, failing to respond even though there are opportunities for them to help themselves by avoiding unpleasant circumstances or by gaining positive rewards. The more people perceive events as uncontrollable and unpredictable, the more stress they experience and the less hope they have about making changes in their lives.

The challenges returnees face are often a result of the nature of the war, and how the issue of relations between the different ethnic groups are still portrayed by the media. As a refugee we interviewed in one of our field studies formulates:

> Today, it seems that there are lot of ethnic tensions, similar to 1992. Media plays a role in igniting these tensions by using hate speech which affect people every day. There are many positive stories in BiH, but media does not focus on the positive, instead whenever there is some positive improvement it seems that media focus is to discourage it directly or indirectly.

There also seems to be a clear understanding of the varied role of different stakeholders in this process. In the words of another informant: 'Private media seem to try to report on positive stories, while the public media usually only reports political and sensational stories; they do not represent the real story.' Our informants generally felt it was very important that people had the opportunity to meet in person, hear each other's sufferings and open up dialogue and communication: 'People in BiH deserve to hear the truth; not the truth driven by politics but the real truth from real actors and individuals who lived through those unfortunate experiences.'

Given the informants' personal experiences, losses and traumas, and the context of a deeply troubled country where mistrust, hatred and prejudice still run deep, with a political climate that is extremely unfavourable for any socio-economic progress, it comes as no surprise that this sense of learned helplessness is constantly being reinforced. Accustomed to a 'culture of donations', returnees often remain inactive and in a constant state of need that, as they see it, can only be met by external assistance. The very high rate of unemployment among returnees, particularly women, mostly due to lack of employment opportunities in their respective communities, and the general lack of skills in demand in the Bosnian labour market, also hinders successful reintegration which in turn further aggravates this feeling of helplessness.

As one of our informants aptly formulated with regard to a project that demonstrated a high level of participation, the exact opposite of what fosters dependency:

> The projects are important not only because they have specific objectives but also because after the war people got used to donations and help from others. Now they do not have working habits or will to engage. Older people remember working actions which took place after World War II and today these types of actions are again very important, especially in an under-developed country such as this one ... Local government has to be included much more. Of course, donations are important but local governance involvement is crucial.

Our accumulated impressions on these issues may be summarized through the following key points.

- Initial forms of providing aid during and especially just after the war were often provided without regard to the context and the actual needs of beneficiaries. That occasionally led to laughable situations, when beneficiaries received city buses regardless of their claims and needs (which may have been related to healthcare, for instance); or hundreds of cows adapted to the abundant grass of the Dutch flatlands were transported to mountainous regions of Eastern Bosnia, with conditions that proved to be in practical terms unviable (the cows starved or fell on the slopes when searching for more grass).

- Donor agencies occasionally intentionally ignored contextual conditions and factors and thus potentially inflated the portrayal of the problem, for example by ignoring the fact that positive legislative solutions for a given problem existed before the war; taking past positive experiences into account would make it more difficult to find a rationale for providing that sort of assistance when seeking money from their benefactors. To quote one such individual prominent benefactor and donor, who will remain unnamed, 'I pay attention when I earn money, not so much when I spend it'.
- Over the two decades since the war, the donors did show a tendency to increase communication with the authorities and the target groups—the refugees or (later) internally displaced persons as beneficiaries—either directly or through subcontracted implementing agencies in the field, which sometimes had the status of partners.
- There has been an increase in providing forms of aid that were meant to be taken over by the BiH authorities as durable and self-sustainable solutions, and oriented towards concrete outcomes, although there were no mechanisms to ensure that actually happens.
- Involvement of consultants in collecting data on the target groups is often constrained and even guided by the contracting agencies. Since regions with refugees and internally displaced persons often treat science as a luxury, the funding and access to migrants are typically in the hands of those providing aid. BiH is no exception. When the consultants offer suggestions and recommendations, they are bound to be in line with the expectations of the contractor, to improve the odds of future engagements. The chances of addressing issues outside the focus of interest of the donor are therefore minimized.
- Migrants are often at the intersection of aid provision by various donors. It quite often happens that the beneficiary is completely oblivious of what agency actually provides a particular kind of support. We faced a situation in which a recipient of direct financial aid believed it was the bank that gave him his monthly stipends.
- Migrants do invariably realize that the assistance they receive is contingent on their poor living conditions. And this sometimes applies equally to families with highly skilled members, relatives abroad, or otherwise able to secure lucrative positions. It is therefore no surprise to encounter instances of households that undoubtedly have considerable income, although there is no way to trace its source. Even when informants own up to moonlighting, this is clearly an acknowledgment of *snalaženje*, an undeniable fact of survival, even a whole way of life, in post-war BiH. Alternative forms of placement meant to serve as a way out of the collective accommodation centre thus easily become weekend houses, or simply one of several items of real estate.
- With time passing since the war, it becomes increasingly difficult to discern what kind of destiny would be typical of this group, compared to the general population. Identification with the group seems to have a primary role to justify a recipient

> role with the compatible and complementary role of the donor. A migrant is a migrant to an assistance provider, but to others, and increasingly with time, they may simply be colleagues, students or friends. As in the humorous definition of a sadist as a person who is kind to a masochist, donors and their beneficiaries apparently have a dynamic relationship of providing each other's *raison d'être.*

So what is the source of the capacity for *snalaženje*? How can it be that even those who do not find themselves on the side of those who get an above-average income, and simply keep on living the lives of forced migrants for years, in most cases do not develop significant problems in the realm of mental health? In instances where we did encounter problems that could have been proved to be clinically relevant, these issues could typically be attributable to causes equally probable in the general population. One of our informants had moved 43 times since the beginning of war. Yet one of us (BS) moved over 20 times in 25 years. Such circumstances offer proof of resilience and hardiness as much as they pose the risk of sinking under pressure.

Discussion

In brief, what has been observed and said about the experiences of Bosnian migrants and internally displaced persons does not deviate significantly from the image portrayed by Barbara Harrell-Bond, who explicitly ties this phenomenon to the 'manner in which relief is given and the supplicatory role which the refugee is forced to assume in the initial period of an emergency',[33] as well as the key phenomena described by her precursors, Frantz Fanon[34] and Pearse and Stiefel.[35]

A number of authors have studied these phenomena over recent decades, emphasizing that '[t]his can happen especially in those camps that reproduce the authoritarian regimes from which the refugees escaped, possibly reproduced by the militarist approach of some relief agencies or peacekeepers',[36] concluding that becoming 'over-institutionalized' causes refugees to lose self-confidence and become numb and apathetic.[37]

> To make matters worse, after imposing learned dependency, donor agencies tend to complain about the dependent and inert behavior of the refugees. In combination with previous traumatic stress and other risk factors such as marginalization, discrimination, poverty, poor physical health, collapse of networks and acculturative stress, this process easily leads to secondary traumatization.[38]

The crucial role of learned helplessness in the dependency syndrome seems obvious: 'when self-reliance is attached to the refugees' recognition that the soil presently beneath their feet is bequeathed by people believing their return home is not imminent'.[39]

But the picture is not so grim, after all. As a refugee pointed out to Barbara Harrell-Bond: 'The fact that our status has changed does not mean that our abilities have gone down.'[40] Furthermore, it appears that people in the lowest of the social classes generally seem to score about twice as high on the wise reasoning scale as those with higher education and regular (high) income.[41] Perhaps poverty, or even more the shifts between classes of socio-economic status implied by the curse mentioned at the beginning of this text, teach a person about the fickle and transitory nature of reality, bringing about a sort of wisdom, even phronesis in what may matter more—one's personal relationships. This testifies to the stance that it may actually be the impact of adversity that awakens certain qualities and skills in people, whether we attribute them to the effects of trauma or some other phenomenon stemming from humanistic psychology.

According to theories of perceived control,

> in order to feel that one has control of one's outcomes one must make two simultaneous judgments. First, the individual must believe that that the situation can be changed if certain actions are taken (that outcomes are contingent on behavior) [and s]econd, the individual must also believe that the relevant agent (oneself or the group) is able to execute the behaviors necessary to produce the desired change.[42]

This appears to offer a way out of the mechanism of learned helplessness, wherein helplessness is a 'disaster for organisms capable of learning that they are helpless'.[43] What we know very little about is how these choices are made by those who thrive despite the adversities of forced migration.

It was again Harrell-Bond who clearly pointed out that consultants and agencies have no accountability when it comes to the impact of the programmes they conceive.[44] As it does not seem plausible to bring about a significant change in the practices of assistance-providing agencies worldwide when it comes to facilitating dependency vs. participation, it may prove fruitful to explore where the limits of dependency are, and when and where the residual capacities manifest themselves to ensure survival. Could it be that dependency is a mere facet of adaptation and thriftiness in allocating one's own capacities and resources?

Conclusions and recommendations

Future studies need to address the issue of the context, funding and geographical setting of data collection. Programme-oriented studies, driven by donors, are unlikely to reveal new facets of the relationship between mental health and forced migration.

The ways in which humanitarian assistance is provided, and how migrants are welcomed in their new environments, may foster what is known as a dependency syndrome, but the full understanding of the outcomes of these processes can be reached only when previous conditions are taken into account, including the value system and the general mentality of people.

Factors known to contribute to the dependency syndrome may superficially reduce the repertoire of behaviours among the refugees and internally displaced persons, with which they could otherwise significantly secure better means for themselves, but that effect does not necessarily permanently deprive them of such capacities. What needs to be addressed in future research is whether such effects are observable primarily on a population scale, while the individual effects remain variable.

Research on migrants is typically focused on the personal level of functioning, such as the indicators of mental health. What may be needed is to address how the experience of migration is affecting more distant processes, especially across generations. Is this experience influencing the system of values and the kinds of expectations that child(ren of) migrants incorporated into their personality?

Bosnians sometimes use the expression that someone has behaved like 'a pumpkin in the water' (in Bosnian: '*kao tikva niz vodu*'), often portraying extreme passivity or being resigned to one's fate. However, we would like to point out that the metaphor may have additional connotations with regard to resilience: the stream may take one in unwanted directions, but the individual capacities remain intact. The pumpkin does not sink; and once it reaches the shore, it is still largely unchanged, despite its turbulent experience. Therefore the metaphor of the floating pumpkin may also be appropriate to depict such a combination of unfavourable social conditions and processes and a retained hardiness and resilience of the individual.

Dependency may simply be a way of coping, contingent on the duration of paternalistic forms of assistance. At this point we know very little of the long-term effects of such phenomena. Processes analogous to post-traumatic development may in effect be increasing the potential of the individual to develop a form of understanding the world around them, a sort of personal wisdom. A pumpkin can also actually help someone else stay afloat. Is there not room for positive psychology in research on people who have undergone involuntary migrations?

Notes

1. I. Cvitković, 'National and confessional image of Bosnia and Herzegovina', *Occasional Papers on Religion in Eastern Europe*, 36(5), 2016, p. 3; I. Cvitković (ed.), *Demographic and Ethnic Changes in BiH*, Academy of Sciences and Arts of Bosnia and Herzegovina, Special Editions, Volume CLXXII, Department of Social Sciences, Volume 9 (in Bosnian).
2. F. Bieber, *Post-War Bosnia. Ethnicity, Inequality and Public Sector Governance*, Palgrave Macmillan, New York, 2006, p. 14.
3. M. Eisenbruch, 'Cross-cultural aspects of bereavement. I: a conceptual framework for comparative analysis', *Culture, Medicine and Psychiatry*, 8(3), 1984, pp. 283–309.
4. M. Parkes, 'Psycho-social transitions: a field for study', *Social Science & Medicine*, 5(2), 1971, pp. 101–115.
5. B. Luthar and M. Pušnik, 'Part I Introduction: the lure of utopia: socialist everyday spaces', in B. Luthar and M. Pušnik (eds), *Remembering Utopia: The Culture of Everyday Life in Socialist Yugoslavia*, New Academia Publishing, Washington, DC, 2010, p. 17.
6. M. Juergensmeyer, 'Nationalism and religion', in A.R. Segal (ed.), *The Blackwell Companion to the Study of Religion*, Blackwell Publishing, London, 2006, p. 361.
7. M. Velikonja, 'Sharpened minds: religious and mythological factors in the creation of the national identities in Bosnia-Herzegovina', in C. Hawkesworth, M. Heppell and H. Norris (eds), *Religious Quest and National Identity in the Balkans*, Palgrave Macmillan, New York, 2009, p. 219.
8. D. Guzina, 'Dilemmas of nation-building and citizenship in Dayton Bosnia', *National Identities*, 9(3), 2007, pp. 217–234.
9. A. Delić, M. Hasanović, E. Avdibegović, A. Dimitrijević, C. Hancheva, C. Scher, T. Stefanović-Stanojević, A. Streeck-Fischer and A. Hamburger, 'Academic model of trauma healing in post-war societies', *Acta Medica Academica*, 43(1), 2014, p. 79.
10. A. Bogdanić, 'Secularizing the ideology of nationalism and nationalizing the ideology of secularism: a refeudalization of discourse in Bosnia and Herzegovina', *Social Identities*, 20 (1), 2014, p. 7.
11. Guzina, op. cit., p. 231.

12. *Comparative Analysis on Access to Rights of Refugees and Displaced Persons*, Ministry of Human Rights and Refugees of Bosnia and Herzegovina, December 2005.
13. *Report on the Realisation of the Revised Strategy of Bosnia and Herzegovina with regard to the Implementation of Annex VII of the Dayton Peace Accord for the Year 2016*, Ministry of Human Rights and Refugees of Bosnia and Herzegovina, April 2017.
14. B. Babic, 'The migration–development nexus in Bosnia and Herzegovina: Center for Local Development and Diaspora seen "from below"', *Southeast European and Black Sea Studies*, 13(2), 2013, p. 213.
15. S. Jansen, 'Troubled locations: return, the life course, and transformations of "home" in Bosnia-Herzegovina', *Focaal*, 49, 2007, p. 26.
16. *The CIA World Factbook 2017*, Skyhorse Publishing, New York, 2016, p. 106.
17. L. Huttunen, '"Home" and ethnicity in the context of war: hesitant diasporas of Bosnian refugees', *European Journal of Cultural Studies*, 8(2), 2005, pp. 177–195; Jansen, op. cit.; S. Jansen and S. Löfving, 'Introduction: movement, violence, and the making of home', *Focaal*, 49, 2007, pp. 3–14; L. Huttunen, 'Sedentary policies and transnational relations: a "non-sustainable" case of return to Bosnia', *Journal of Refugee Studies*, 23(1), 2010, pp. 41–61; A. H. Stefansson, 'Coffee after cleansing? Co-existence, co-operation, and communication in post-conflict Bosnia and Herzegovina', *Focaal*, 57, 2010, pp. 62–76; L. Huttunen, 'From individual grief to a shared history of the Bosnian war: voice, audience, and the political in psychotherapeutic practices with refugees', *Focaal*, 68, 2014, pp. 91–104.
18. M. Hasanović, 'Neuroticism and posttraumatic stress disorder in Bosnian internally displaced and refugee adolescents from three different regions after the 1992–1995 Bosnia-Herzegovina war', *Pedijatrija Danas: Pediatrics Today*, 8(2), 2012, pp. 100–113; M. Hasanović, O. Sinanović and S. Pavlović, 'Acculturation and psychological problems of adolescents from Bosnia and Herzegovina during exile and repatriation', *Croatian Medical Journal*, 46(1), 2005, pp. 105–115.
19. A. Hasanagić, ['A study of psychological well-being among women forced migrants in Bosnia and Herzegovina'], in S. Porobić (ed.), *Rodna studija prisilne migracije i psihosocijalnog dobrostanja u Bosni i Hercegovini*, Fondacija za osnaživanje žena BiH, Sarajevo, 2016, pp. 31–68 (in Bosnian).
20. E. Avdibegović and O. Sinanović, 'Consequences of domestic violence on women's mental health in Bosnia and Herzegovina', *Croatian Medical Journal*, 47(5), 2006 pp. 730–741; D. Baraković and E. Avdibegović, 'Psychological disturbance in women with experience of missing family members' (20th European Congress of Psychiatry, P-1452), *European Psychiatry*, 27, 2012, p. 1.
21. State law postulates that 'A displaced person is a citizen of Bosnia and Herzegovina, residing within Bosnia and Herzegovina, who has been expelled from his/her habitual residence as the consequence of the conflict, or left her/his habitual residence, after 30 April 1991, owing to a well-founded fear of being persecuted for reasons of race, religion, nationality, membership to a social group or political opinion, and who is neither able to return in safety and with dignity to his/her former habitual residence nor has voluntarily decided to take up permanent residence elsewhere'.
22. The new provision implies that one can receive compensation for the destroyed property even one does not sign up for return or return voluntarily.
23. A. Cavalli, 'Reflections on political culture and the "Italian national character"', *Daedalus*, 130(3), 2001, p. 126.
24. R. Perin and F. Sturino (eds), *Arrangiarsi: The Italian Immigration Experience in Canada*, Guernica, Montréal, 1989.
25. P. Weiss Fagen, 'Refugees and IDPs after conflict: why they do not go home. Special report 268', US Institute of Peace, Washington, DC, 2011, p. 4.
26. I. Maček, *Sarajevo Under Siege: Anthropology in Wartime*, University of Pennsylvania Press, Philadelphia, 2009.

27. M. Povrzanovic Frykman, 'Sensitive objects of humanitarian aid: corporeal memories and affective continuities', in J. Frykman and M. Povrzanovic Frykman (eds), *Sensitive Objects: Affect and Material Culture*, Nordic Academic Press, Lund, 2016, p. 100.
28. R. K. Papadopoulos, 'Refugees, trauma and adversity-activated development', *European Journal of Psychotherapy and Counselling*, 9(3), 2007, p. 302.
29. B. E. Harell-Bond, *Imposing Aid: Emergency Assistance to Refugees*, Oxford University Press, Oxford, 1986.
30. M. E. Seligman, *Helplessness: On Depression, Development, and Death*, W. H. Freeman and Company, San Francisco, 1975.
31. A. Ovčina, 'Position of internally displaced women in today's Bosnia and Herzegovina. a study in situational gaps and good practices related to access to employment', Open Society Fund, 2013 <http://www.osfbih.org.ba/images/Prog_docs/PDFP/pdfp_13/ENG_S_77_Adnan_Ovcina.pdf> (accessed 6 January 2018).
32. N. Bosankic, 'Ethnography with displaced women from Drinjevac', in S. Porobic (ed.), *Gender Study of Forced Migration and Psychosocial Wellbeing in Bosnia and Herzegovina*, Foundation for Women's Empowerment, Sarajevo, 2016, pp. 69–98; N. Bosankić, E. Mešić and B. Šošić, *Impact Study of Social Housing in Bosnia and Herzegovina*, Catholic Relief Service (CRS), Sarajevo, 2013; N. Bosankić and E. Mešić, *Social Housing in Bosnia and Herzegovina Project—Final Evaluation Report*, Catholic Relief Service (CRS), Sarajevo, 2011; B. Šošić and N. Bosankić, *Social Inclusion of Women Returnees: Gender Sensitive Situation Analysis*, TPO Foundation, Sarajevo, 2010.
33. Harrell-Bond, op. cit.
34. F. Fanon, *The Wretched of the Earth*, translated [from the French] by Richard Philcox; introductions by J.-P. Sartre and H. K. Bhabha, Grove Press, New York, 1961/2004.
35. A. Pearse and M. Stiefel, 'Inquiry into participation—a research approach', UNRISD/79/C.14, Geneva, May 1979.
36. J. T. V. M. de Jong, 'Public mental health, traumatic stress and human rights violations in low-income countries', in J. de Jong (ed.), *Trauma, War, and Violence: Public Mental Health in Socio-Cultural Context*, Kluwer Academic Publishers, New York, 2002, p. 76.
37. U. von Buchwald, 'Refugee dependency: origins and consequences', in A. J. Marsella, T. Bornemann, S. Ekblad and J. Orley (eds), *Amidst Peril and Pain: The Mental Health and Well-Being of the World's Refugees*, American Psychological Association, Washington, DC, 1994, pp. 229–237.
38. de Jong, op. cit., p. 76.
39. N. Baron, 'Community based psychosocial and mental health services for southern Sudanese refugees in long term exile in Uganda', in J. de Jong (ed.), *Trauma, War, and Violence: Public Mental Health in Socio-Cultural Context*, Kluwer Academic Publishers, New York, 2002, p. 162.
40. Harrell-Bond, op. cit., p. 12.
41. J. P. Brienza, and I. Grossmann, 'Social class and wise reasoning about interpersonal conflicts across regions, persons and situations', *Proceedings Royal Society B*, 284(1869), 2017, p. 20171870.
42. S. C. Wright and L. R. Tropp, 'Collective action in response to disadvantage: intergroup perceptions, social identification, and social change', in I. Walker and H. J. Smith (eds), *Relative Deprivation: Specification, Development, and Integration*, Cambridge University Press, Cambridge. 2002, p. 221.
43. Seligman, op. cit., p. 44.
44. Harrell-Bond, op. cit., p. 14.

Disclosure statement

No potential conflict of interest was reported by the authors.

The Serbian Diaspora's Political Views: A Study on the 2017 Serbian Presidential Election

Tena Prelec

ABSTRACT

Members of the Serbian diaspora, especially those vocal in representing their home country's views, have previously been regarded as leaning towards nationalistic stances. The results of recent elections, however, indicate that today a rather different situation might be present. This study confirms this indication. The analysis sheds light on the political views of Serbian citizens abroad, discussing the results of two surveys conducted around the 2017 Serbian presidential elections. Three main sets of findings concerning the attitudes of the 'active diaspora' (diaspora members who took part, or wanted to take part, in the voting process) are outlined. First, there is a decisive rejection of the course the country has taken. Second, it is shown that concerns about governance top by far geopolitical issues. Finally, respondents have highlighted a wide-ranging set of issues concerning the voting procedure, which have hindered their ability to participate in the vote. Here they are used as a basis to formulate recommendations for a better electoral process for the diaspora. The overall findings indicate that Serbia's active diaspora might have the potential to participate more vigorously in the political life of the country and that, if emboldened, its weight in future elections could be much higher.

Introduction

Approximately one in ten Serbs live abroad: the latest data from the International Organization for Migration[1] put the 'stock' of outward migration from Serbia at 964,585, noting that 9.8% of all citizens of Serbia live outside their country of origin. Other unofficial estimates quote an even higher number, putting the number of Serbian citizens living abroad at totals ranging between 2 and 3 million.[2] This is, in any case, a large contingent, when compared to a population of about 7 million living in Serbia.[3] Out of these, only 9523 Serbs abroad made use of their right to vote in the 2017 presidential election. Why this gap? What are the current political views of the diaspora, and what can be done to further stimulate their engagement? This paper addresses these questions through a study carried out during the 2017 presidential vote. It will be argued that, while a level of disaffection with the political processes in the country of origin is only natural for citizens living abroad, this large gap is likely to be partially accounted for by the difficulties experienced during the registration process.

Concerning the political stances of the diaspora, this study highlights that—contrary to previously held beliefs—a majority of the politically active members of the diaspora (i.e. those who intended to take part in the voting process) do not show great interest in topics that are usually linked to nationalistic stances, while expressing the highest concerns over governance issues. Pooling responses from over 300 members of the Serbian diaspora, the study furthermore suggests a number of practical actions which can be carried out in order to increase the political participation of the diaspora. An open question remains as to the presence of the political will to implement such relatively simple changes, which could have a considerable impact on future elections.

Literature review: previous studies on diaspora's political views and the need to move the research agenda forward

Members of diasporas have long been perceived to have more extreme views than the locals—i.e. those who have stayed in the home country—after a war. So far, migration dynamics from the Balkan countries have mostly been looked at through this post-conflict lens. That diaspora members hold more radical political views is indeed a widespread assumption in the literature on migration, and well beyond South-East Europe,[4] though it is not one that has been supported by solid empirical findings. The explanations upon which this postulation rests range from the need to maintain a collective identity in the face of a new, alien territory[5] or to preserve myths and political ideologies;[6] to the radicalization that occurs in neighbouring countries, where 'temporary' camps may become permanent;[7] or to the fact that locals may be incentivized to compromise and forge a common life,[8] while migrants would become polarized because of the lack of contact with an adversary.[9] In a nutshell, the assumption is that 'remembering a conquered or oppressed home is one of the duties of emigrants; the problem is that exile can freeze conviction at the moment of departure. Once in exile, groups fail to evolve'.[10]

The diasporas from the Western Balkans have very often been associated with precisely these dynamics, something that has been accentuated by the disproportionate attention that the scholarship on the region has given to war and conflict. And while these insights should be kept in mind, this line of inquiry urgently needs to be complemented by fresh thinking. A recent study by Jonathan Hall, comparing the views and attitudes of Bosnians who have stayed in their country with those who left in the 1990s, found out that migrants do not necessarily hold more extreme views than those who stayed.[11] Analysing data from simultaneous surveys conducted in the two countries, Hall's research showed that, under certain conditions, migrants can benefit from opportunities for reconciliation and assimilation offered by the recipient country, and that in such cases their views can indeed become more conciliatory than those of their peers who stayed in the home country. Another recent study[12] highlighted the positive experience of Bosnian refugees in three countries out of the four considered, signalling once again the importance of investigating the effect of the policies of the recipient countries on migrants' attitudes.[13]

This paper confirms that this novel trajectory is an important one, and worth exploring further. Its findings corroborate the need to problematize this key assumption in the literature on migration and the Western Balkans. It will be shown that a significant part of the Serbian diaspora—specifically, the more highly educated segments of the population who have moved abroad in recent times and who make up the majority of the country's politically active diaspora, i.e. who take an interest in the home country's political and electoral process—do not fit in the above-mentioned categorization of hard-liners unable to evolve their views. Quite the contrary, these individuals belong to a category of people who are unhappy with the autocratic direction taken by their government, and hold diametrically opposed views to those espoused by the country's leadership.

Clearly, the discrepancy between the key assumption discussed above (that migrants hold more extreme views than their peers who stayed in the home country) and the more recent works disproving it, is partially to be sought in a change in migration patterns, which have ostensibly taken a different nuance in more recent times. The lack of economic prospects, and even more the lack of a level playing field in the home country, is making many young people leave the Western Balkans, with very grave consequences at present—and potentially detrimental consequences in the future—for the economic, social and political fabric of those countries. This is something that requires a robust new research agenda. It is thus long overdue to abandon the lens of war and conflict to explain the migration dynamics concerning Balkans.

The 2017 presidential vote and the diaspora

Data collection was carried out in conjunction with the 2017 Serbian presidential election. The vote took place on 2 April,[14] with a hypothetical run-off initially foreseen for two weeks later. The voting process abroad was hasty, and fraught with communication issues and physical barriers. The registration process for the diaspora was announced on 3 March on the website of the Ministry of Foreign Affairs of Serbia[15] and voters had eight days, i.e. up until 11 March, to sign up. Polling stations were opened only in places where at least 100 voters had registered. Postal voting was not allowed, and voters were compelled to cast their ballots in the same place in both rounds: this made it impossible for those travelling to Serbia for Easter holidays to express their preference in a hypothetical run-off, which was scheduled to happen on Easter Sunday—but ended up not taking place due to Aleksandar Vučić's victory in the first round. As a result, only 53 voting stations were opened worldwide, and 11,590 Serbian citizens abroad registered to vote, with merely 9523 people casting their ballots on voting day.[16] No voting stations were opened in a number of key diaspora countries—Australia being just one case in point (see Figure 1). Figure 2 shows the voting stations in Europe in more detail.

The runners and riders and the outcome

A short note about the most prominent of the 11 candidates is in order. Aleksandar Vučić, the overall winner of the election—who was, at the time of the vote, the incumbent Prime Minister and leader of the Serbian Progressive Party—decided to stand for the presidency after internal polling showed that the outgoing President and co-party member, Tomislav Nikolić, would risk losing the contest. A nationalist in his youth, hailing from Vojislav Šešelj's Radical Party and having served as Minister of Information under Slobodan Milošević, Vučić had successfully carved an image of

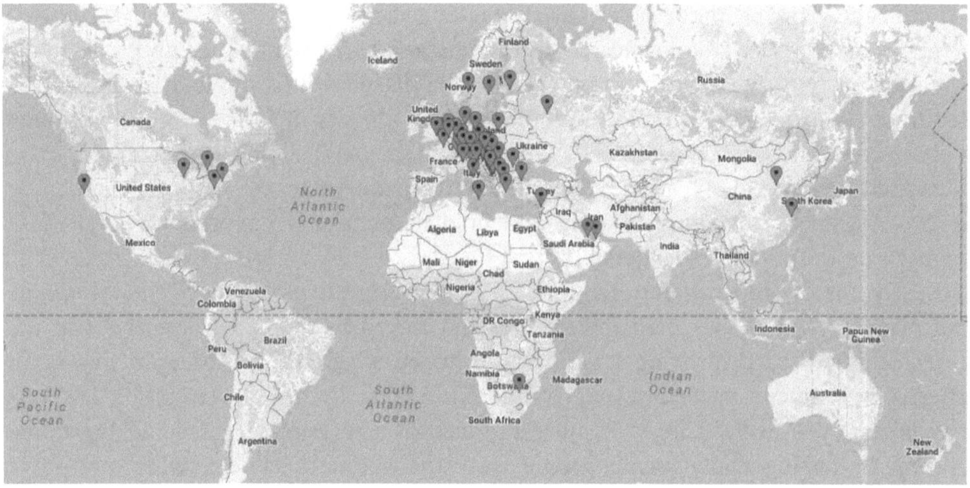

Figure 1. Diaspora voting stations worldwide.

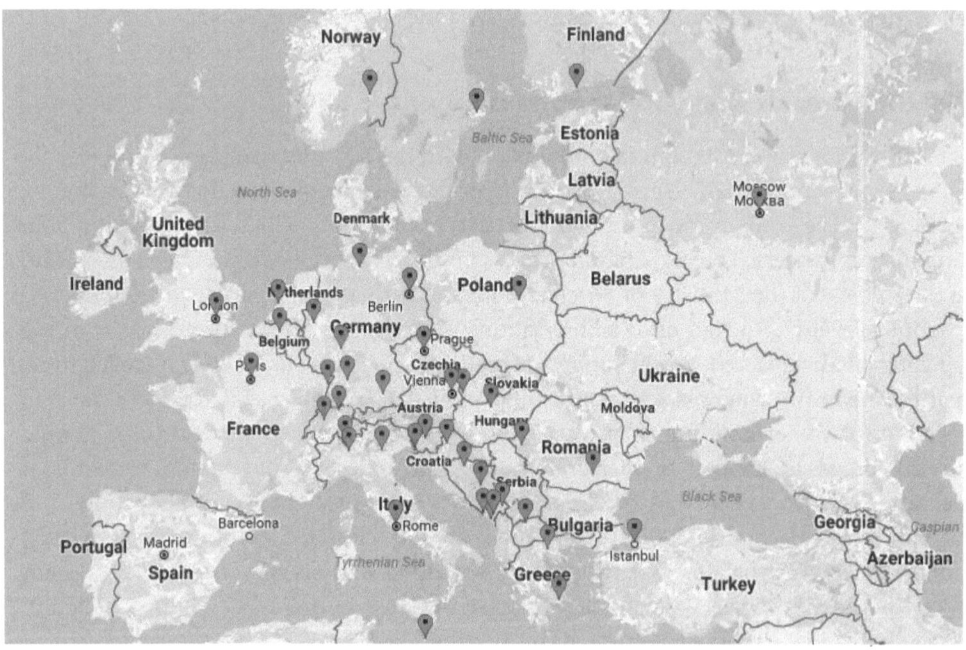

Figure 2. Diaspora voting stations in Europe.
Source: RIK (Serbian Electoral Committee). Compiled by the author.

a man who juggles a pro-European agenda while still representing the national interests of Serbia. His style of government is unmistakeably top-down, concentrating the decision-making in his hands and monopolizing media attention[17]—features that did not change in his passage from Prime Minister to President. He was the clear front-runner from the beginning.

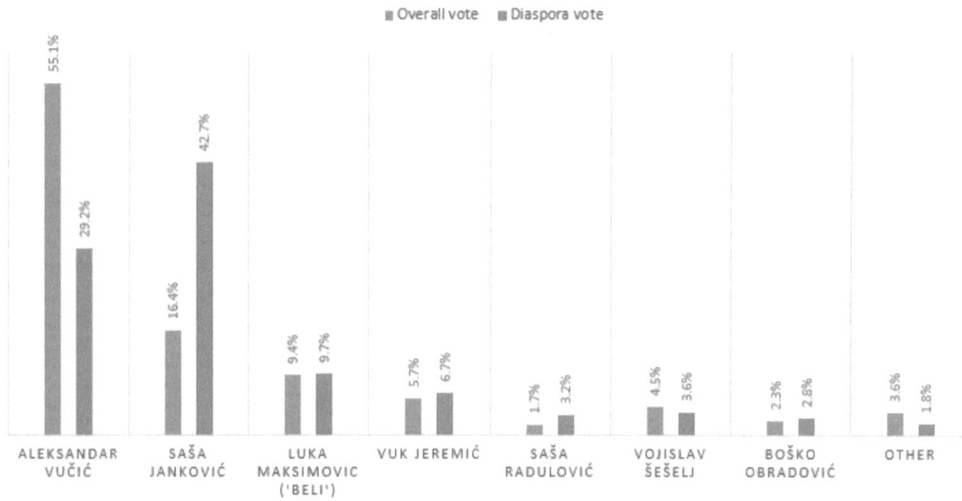

Figure 3. Serbian presidential election outcomes—overall vote and diaspora vote compared.
Source: RIK (Serbian Electoral Committee).

Several new challengers from different strands of the opposition had also thrown their hats in the ring: former Ombudsman Saša Janković, former Foreign Minister and UN General Assembly President Vuk Jeremić, comedian Luka Maksimović (running as the spoof candidate Ljubiša 'Beli' Preletačević, a reference to the infamous practice of politicians frequently changing political parties out of convenience), and Enough is Enough leader Saša Radulović. Other challengers included Vojislav Šešelj of the nationalist Radical Party, and Boško Obradović, who heads the far-right movement 'Dveri'. The campaign was therefore an exceptionally lively one, with large rallies and the parody candidate 'Beli', who entered the contest at a late stage, gaining very significant traction in the last weeks of the campaign and ending in third place.

The result of the diaspora vote, however, indicated a reversed outcome: there was a marked preference for the opposition candidate Saša Janković (Figure 3), while the ballots for Vučić were almost halved in comparison to the overall outcome. Janković represented a broad progressive platform that appealed mostly to city-dwellers.[18] His candidacy pitch highlighted European and liberal values and denounced what he referred to as '*bratija*', literally 'brotherhood', meaning the informal network of favours that, as he maintains, has taken the place of the state.[19] Similar messages were pushed by Saša Radulović, Vuk Jeremić and (in a reverse fashion) by the comedian Maksimović.

The surveys

To understand better the political attitudes of this potentially large but understudied contingent of voters—the Serbian diaspora—I first created a pre-electoral survey (Survey 1), distributed in the week before the vote (25 March–1 April 2017).[20] Survey 1 was shared with individuals and closed diaspora groups, rather than publishing the link more widely. Aside from basic demographic questions, respondents were asked to identify their current country and city of residence; how long ago they moved to Serbia;

what source(s) of information they use when learning about Serbian politics; what were their voting intentions for the 2 April 2017 presidential election and for a hypothetical run-off; what they thought were the voting intentions of their relatives and friends abroad and in Serbia; whether the electoral campaign impacted their opinion of the presidential candidates; whether they were satisfied with the voting procedure for Serbs above and why; what, according to them, the key issues in Serbia today were; and what their overall opinion of the Serbian government was.

A further follow-up survey (Survey 2), focusing on the views of the diaspora on political engagement, was circulated among Survey 1 respondents after the election.[21] This survey collected more in-depth, qualitative responses on their current involvement in political activism; on their propensity to get involved in civic and political movements in (or on) Serbia; on their thoughts about returning to Serbia one day; and on what they see as the role of the Serbian diaspora and whether the diaspora should take part in the economic, political and social developments at home.

Considering the groups and individuals from whom the surveys started to be disseminated, I expected the immediate snowballing effect to have reached especially the more highly educated strata of the Serbian diaspora population. The education level of the respondents was indeed very high, with almost 70% in possession of a Master's degree or a PhD (Figure 4).

I therefore attempted to correct for these foreseen biases by pro-actively contacting respondents who are part of more traditional or nationalistic-leaning groups, reaching out to communities linked to the Serbian Church, and even by flagging the survey up to diaspora groups of the Serbian Progressive Party. Even so, the results obtained in terms of voting preferences (which have largely matched the eventual outcome), show a widespread rejection of the current government, and of Aleksandar Vučić in particular, among the diaspora members surveyed.[22]

The surveys achieved good geographic coverage, collecting responses from respondents living in 47 countries; and the overall sample ($N = 322$) is significant, if set against

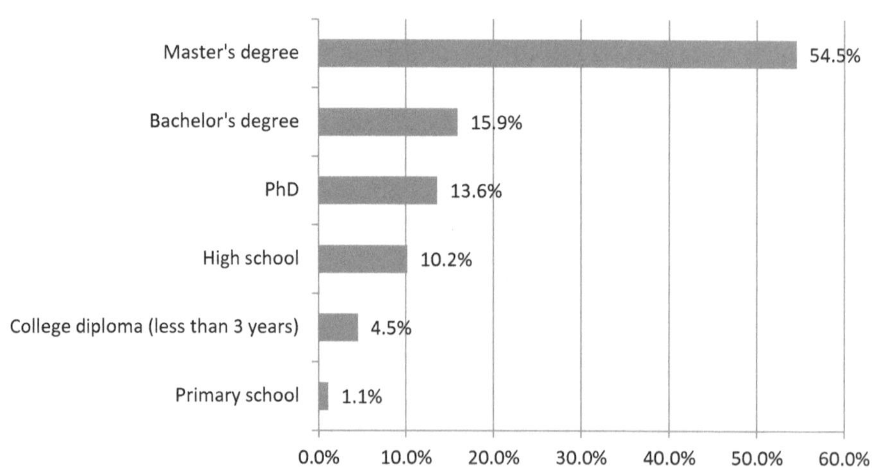

Figure 4. Highest level of education completed.
Source: Survey 2.

the number of registered voters (11,590). Out of the 47 countries, the main ones were as follows (with the number of respondents in brackets): Germany (52), UK (37), Austria (24), Italy (20), UAE (19), USA (17), France (15), Norway (15), Switzerland (14), the Netherlands (12), Sweden (9) and Canada (8).[23]

Opposition to the government

The most striking response regards the opinion of the current Serbian government: over 71% of respondents chose to tick the 'very negative' box (Figure 5), and a further 21.1% opted for 'negative'. These stark figures were echoed by verbal entries in the questionnaire surveys; some examples follow.[24]

First, only a few respondents defended Aleksandar Vučić's work:

> I started following Serbian politics with the rise of Vučić. He is the only candidate that I can follow because he is discussing relevant issues, the economy!!! Everyone else is always involved in some kind of name calling and revolution.

Or expressed mixed opinions:

> Vučić gets my points for understanding the importance of both Germany/the West and Russia. One should not be forsaken for the other and a good politician should be able to play both sides. He understands the importance of PR and media for his own gains and the country's image abroad. He is a better politician than all of the other candidates but he has become a modern dictator, oppressing people by controlling the media.

But most of the respondents expressed a strong desire for renewal and markedly negative opinions of the government, as the following typical entries attest:

> We need to build a new system. A new society. We do not have time nor strength for the [further] implementation of something which is deviant in substance—the result will always be the same.

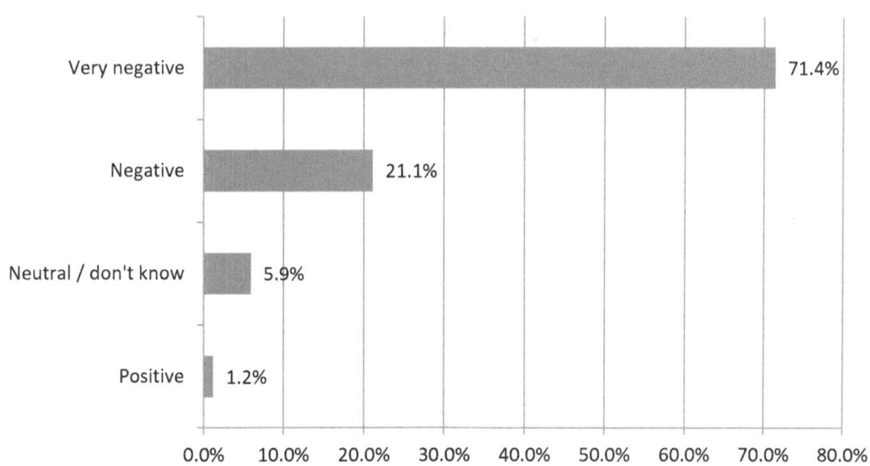

Figure 5. What is your overall opinion of the Serbian government at present?
Source: Survey 1.

> There are many of us who would like to come back and invest in Serbia. To do so, we need strong institutions and a state that works. Most of us have left because we felt like second-rate citizens in our own country. Let's start from a radical change of the taxation system and the judiciary.
>
> Dictatorship. The Serbian Third Reich is upon us if we do not bring Vučić down.
>
> The moment when the current government came to power has been decisive in my decision to leave Serbia. I realized then that I cannot stay in my own country.
>
> Particracy and autocracy are destroying us.
>
> [The government is] criminal, totalitarian, incompetent, immoral, shameless, corrupt, not fit for purpose.

A new cleavage?

The above-described desire for a radical renewal, and opposition to the current course taken by the Serbian political scene is not entirely unusual: it found an immediate echo in the almost month-long protests on the streets of several Serbian cities that followed the 2 April election.[25] Survey 1 respondents were asked to indicate who they thought their friends abroad and those in Serbia were going to vote for. It was expected that a much higher number of responses in favour of Aleksandar Vučić would be ticked in this case, but this was only marginally so: 24 out of 322 respondents thought that their acquaintances in Serbia were going to vote for Vučić, seven that their friends abroad would do so. This seems to point to the fact that members of the active diaspora are connected with like-minded people at home. The election has therefore exposed the increasingly clearer profiling of two distinct groups within Serbian society—one outward facing that is critical of the government, and a more traditional and rural one—interacting very little with one another. One of the key discriminants may well be the media: while television and print are the main sources of information at home, a total of 93% of respondents indicated that their preferred channel of information about Serbian politics is the web (Figure 6), with 47.1% opting for social media and 45.3% choosing online publications.

Concerns: governance over geopolitics

The survey responses shed further light on the main concerns harboured by the active diaspora. In contrast to previously-held beliefs on the nationalistic stances of the Serbian diaspora actively representing Serbia's voice abroad,[26] these responses indicate that the trend has now changed. The issues of Kosovo, of Serbia's relations with Russia and even with the EU, are lowest on the list. Concerns about governance, on the contrary, are highest: corruption takes the first spot, followed by the judiciary, media freedom, autocracy, education, access to the labour market and economic inequality. As Figure 7 shows, all the above issues are highlighted by at least 70% of the survey respondents, ranging up to 88% on the issue of corruption. All other issues polled responses of less than 50%.

Barriers to the vote

Many of the respondents lamented that certain aspects of the voting procedure for the diaspora made it difficult for them to participate in the process, and a great many

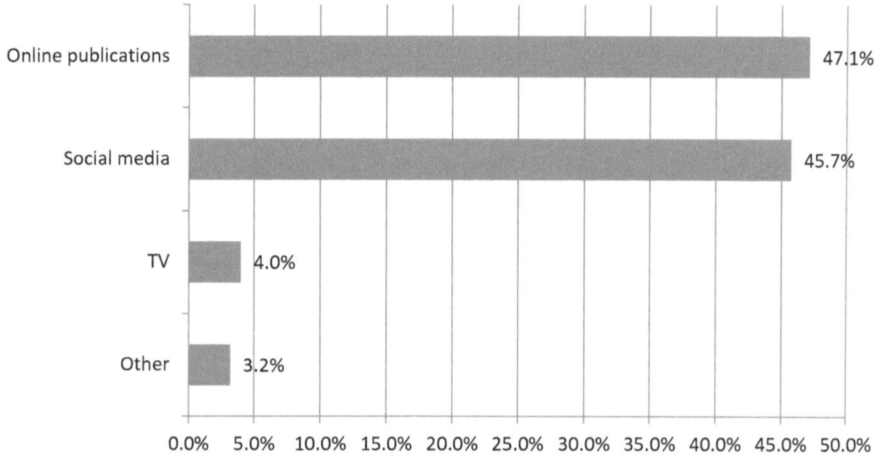

Figure 6. Where do you predominantly get your information on Serbian politics from?
Source: Survey 1.

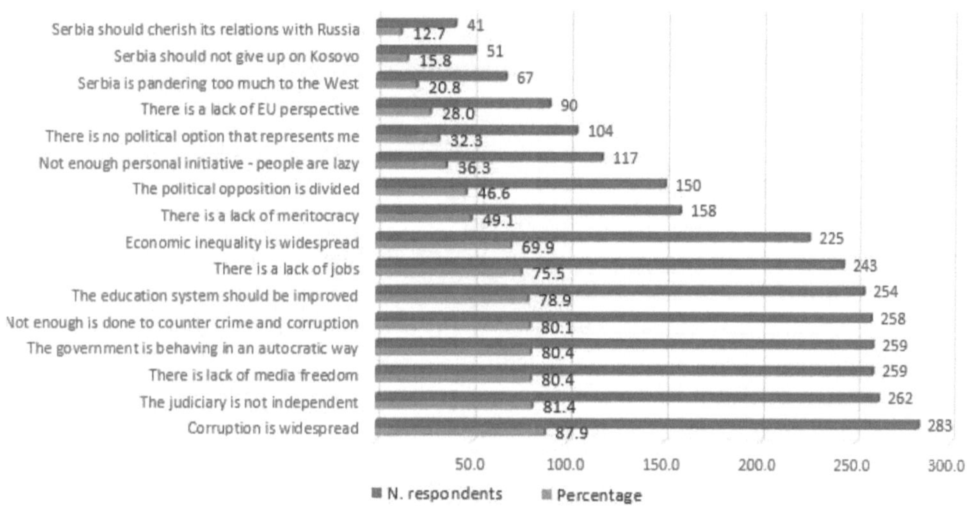

Figure 7. What are some of the key issues in Serbia today?
Source: Survey 1.

complained that they were unable to vote, despite a desire to do so. The registration itself, while not entirely straightforward,[27] did not present excessive difficulty.[28] The most widespread problems however were: (1) *short notice*: voters abroad were given only one week to register for the vote; (2) *communication*: effective communication was lacking—many citizens were not informed or were informed late; (3) *scarcity of voting stations*: the rule that at least 100 voters should have registered to open a voting station meant that in many cases people who lived far away from the main cities could not vote, and in many countries no voting stations were present whatsoever; and (4) *IT issues* (server failures) in embassies and consulates that have jeopardized the registration of voters in more than one city. Furthermore, the *need to vote in the same place in both rounds* (with a hypothetical run-off taking place on Easter day) would have precluded the vote for all those who had travel plans for the Easter holidays. The experience of Ugljesa Grušić, a law lecturer at University College London (UK), is illuminating in this sense:[29]

> I will be in Serbia during the first round and in London during the second round of the election. I emailed the embassy to ask whether I could register to vote in London in the run-off. The embassy replied that they received a 'notification' from the Ministry of Foreign Affairs stating that it is not possible to change the voting location after the first round. I replied to this email quoting Article 23 of the Law on a Single Electoral Roll, the Constitution and the European human rights instruments. I asked them to send me the text of the instruction and to justify their refusal to allow me to exercise my voting right. I received the following reply (my translation): 'Unfortunately, we cannot send you the original text of the notification that we received from the Ministry of Foreign Affairs, since it is part of official communication between the Ministry and the Embassy, but we posted this information on the website'. Needless to say, I felt infuriated and totally powerless when I saw this reply. I feel that there is nothing I can do about this. Serbia is at the moment a country where the constitution and the laws can be suspended by a 'notification' issued by a Government ministry, which notification no one is able to see.

Going forward: how can Serbia empower its diaspora?

How to right these wrongs, and improve the diaspora's political participation? Responses collected through Survey 2 indicate four main ways to achieve this. It should be noted that not all respondents were of the opinion that increased political participation of members of the diaspora should be pursued. In the words of one of them: 'I left my country and have no right to meddle in the fate of the people living there. We are all the victims of our choices.' Most individuals, however, were keen to think of ways to increase their participation, and several of them held a strong opinion in this sense: 'Diaspora members keep Serbia afloat with their remittances. We therefore have full moral right to participate in the political process.'

By far, the most quoted remark concerned the access to the electoral process, highlighting the need to implement a more efficient system—as outlined above. Implementing postal and electronic voting was consistently mentioned as a necessary development to guarantee voting rights for diaspora members. A second aspect raised is the issue of representation in parliament at general elections. Contrary to what takes place in some neighbouring countries (Croatia is a case in point), the diaspora is not given direct representation in parliament. Furthermore, an oft-quoted remark was that embassies and consulates should step up their game in providing a helpful service to Serbs abroad. Complaints were raised mainly in relation to the voting process (informing citizens and guaranteeing their right to vote), but many suggested that the diplomatic bodies should do more to provide occasions for Serbs to get together and organize. To increase the organization of diaspora members, respondents furthermore suggested setting up social media groups, non-profit organizations and websites for the diaspora. It should be noted that, while there is no overarching organization, a few such spontaneous groups already existed, and some were set up specifically for the 2017 presidential elections—such as Glasajmo and Dijaspora Glasa.[30]

Many highlighted that they would be happy to share their skills and knowledge with their compatriots, and a clear majority of respondents said they would at least consider returning to Serbia. Most of them, however, set out quite clear conditions under which they would return. The most quoted motivation was achieving financial freedom, to be able to be free from political influences. As a close second, respondents cited fair competition (meritocracy) as a necessary precondition, making reference to the economic and political situation and advocating the end of corruption and 'particracy'. The concern regarding the society where their children will be growing up was a drawback in this sense: parents with young children quoted this as a reason to stay abroad. Few respondents showed enthusiasm for returning and participating in the life of the country as it stands and, finally, personal motivations (of a non-political nature) were cited as well. It is clear, however, that the political and economic situation in the country—perceived as unjust—was the top concern of the respondents. One individual summarized this by stating: 'As long as corruption and the disregard for rules are present in all spheres of life, I am not thinking about a return to Serbia.'

The insights above show that the political situation back home has the distinct ability to influence the willingness of this class of diaspora voters to remain in their home country, to return to it, and to engage with the political and social developments of the places of origin. This finding is corroborated when assessing the aftermath of the change in the Macedonian government in May 2017, which occurred after a long stalemate and through the engagement of both civil society movements and relevant EU institutions. The grass-root protests were triggered by the discovery, in February 2015, of widespread government-sponsored illegal surveillance, and leaked tapes exposed the extent to which the state was being captured by a corrupt, authoritarian and self-serving elite.[31] Elections in December 2016 resulted in a change in voters' preferences, and eventually in the appointment of a new government. There is, for now, anecdotal evidence that the democratic election of a new reformist government, which represented a change of guard after over 10 years, is directly linked with a shift in the attitude of young people towards migration.[32] After a decade of brain drain, Macedonian youth is now 'not so sure that they want to leave the country any longer', while the new government 'wants to move things forward, but they find it very difficult to find appropriate cadres because of the heavy brain drain that occurred previously'.[33] The brain drain, of course, does not necessarily need to be negative: the impact of returnees on the home country has been shown to be extremely positive.[34] The question is, therefore, how to make best use of the social capital and to feed it back in. It is also a very urgent research question: as the quote above shows, if we have to wait for too long, there will be no one left in the country to carry out a progressive and transparent agenda when the opportunity is presented.

Conclusions

The study outlined in this paper has focused on the political participation of the Serbian diaspora in Serbia's 2017 presidential election. It has concentrated on the views of what has been termed 'active diaspora': individuals living abroad who have taken a clear interest in the home country's electoral process. The research points to the importance of a heavily under-researched topic—the changing views of portions of the Balkan diasporas, the obstacles to their engagement and avenues for their future participation —and it has the ambition to open the door to a number of research questions that are,

I would argue, extremely meaningful. Some of the findings highlight glaring deficiencies in the voting process for Serbs abroad that require prompt and decisive action, and should be of interest to policy-makers.[35]

The main findings can be expressed in three sets. First, *the active diaspora is strongly opposed to authoritarianism* and is widely critical of (Prime Minister and then) President Vučić. The stance of the active diaspora has been shown to be overwhelmingly opposed to the current direction of travel of the political leadership of their country. While this is clear from the result of the diaspora vote itself, the qualitative material collected and analysed above shows the vehemence with which such convictions are held.

Second, *the active diaspora deeply cares about governance issues, disregarding topics linked to international relations and to nationalistic stances.* Governance (corruption, judiciary, press freedom, employment, and education) comes first among the worries of the survey respondents. The very little importance given to issues such as EU accession, relations with Russia and, especially, the question of Kosovo, challenges previously held beliefs that Serbia's diaspora is nationalistic. While it cannot be denied that clubs and associations with more traditional views exist abroad, this study has shown that the politically active members—many of whom have recently moved abroad—are not bothered by these topics. This seems to corroborate the findings by Efendic *et al.*, who remark that individuals with a higher level of education, and with a stable employment status, are less sensitive to the issue of ethnicity.[36] It is also potentially interesting insofar as it diverges from the political views of neighbouring countries' diasporas, namely the Croatian one[37]—which is viewed as holding particularly nationalistic views.[38] Furthermore, most of the surveyed individuals appear to be *connected to like-minded individuals at home.* Respondents indicated that their family and friends in Serbia will be voting for opposition candidates. This suggests the profiling of two groups not along a left–right divide, but rather along a cleavage which is located on a pro- and anti-government axis. Therefore, two groups interacting very little with each other seem to be profiling themselves within Serbian society: one outward-facing that is critical of the government and that includes mostly city-dwellers and a majority of the members of the active diaspora, and a more traditional and rural one. Initial interviews with Serbian activists corroborate this finding,[39] but further research is needed to test these remarks.

Finally, the results presented in this paper show that the political participation of the diaspora is currently severely hindered and has space for growth. The responses collected and analysed recommend a number of uncontroversial 'quick wins' that the government could implement relatively easily starting from the next elections: the implementation of the postal vote; a longer time frame for registrations and an easier registration process; improved communication and better service from embassies and consulates; and making sure there are no IT problems that prevent the registration of voters. Other suggestions are slightly more problematic, but should be taken into account and discussed openly: electronic voting; parliamentary representation for voters abroad; and the full listing of Serbs living abroad.

On the whole, the findings indicate that Serbia's active diaspora has the potential to participate more vigorously in the political life of the country and that, if emboldened, its weight in future elections could be much greater. From a long-term perspective, it is clear that the way to incentivize the participation of the active diaspora in political and economic processes in Serbia is to foster a higher level of equality and meritocracy in Serbian society—an insight that finds corroboration in the shifting attitudes of Macedonians after the democratic change of their government's leadership in 2017, when they are reported to have become less keen to emigrate. The question that remains, however, is whether it is in the interests of Serbia's ruling elite to act upon these insights and allow for a pool of increasingly cosmopolitan and progressively minded citizens to take a more active part in domestic politics. Acting upon the recommendations presented in this paper, i.e. taking concrete steps to remove barriers to the diaspora vote, would be the best way to dispel any doubts that these restrictions to political participation are maintained to further their own electoral scopes.

Notes

1. Source of data is International Organization for Migration <https://www.iom.int/countries/serbia and https://www.iom.int/world-migration> (accessed 23 January 2018).
2. See <https://en.wikipedia.org/wiki/Serb_diaspora> (accessed 23 January 2018).
3. V. Ilić, 'Nevolje sa biračkim spiskom' [Problems with the electoral register], *Peščanik*, 15 April 2017 <http://pescanik.net/nevolje-sa-birackim-spiskom/> (accessed 23 January 2018).
4. See: B. Anderson, 'Long distance nationalism', in B. Anderson (ed.), *The Spectre of Comparisons: Nationalism, Southeast Asia and the World*, Verso, London, 1998, pp. 58–74; M. B. Anderson, *Do No Harm: How Aid Can Support Peace—or War*, Lynne Rienner, Boulder, CO, 1999, p. 18; B. Bigombe, P. Collier and N. Sambanis, 'Policies for building post-conflict peace', *Journal of African Economies*, 9(3), 2000, pp. 333–334; B. Bock-Luna, *The Past in Exile: Serbian Long-distance Nationalism and Identity in the Wake of the Third Balkan War*, Transaction Publishers, New Brunswick, NJ, 2007; P. Collier and A. Hoeffler, 'Greed and grievance in civil war', *Oxford Economic Papers*, 56(4), 2004, pp. 563–595; P. Collier and A. Hoeffler, 'The political economy of secession', in H. Hannum and E. F. Babbitt (eds), *Negotiating Self-Determination*, Lexington Books, Lanham, MD, 2006, pp. 37–59; G. Golan and A. Gal, 'Globalisation and the transformation of conflict', in B. W. Dayton and L. Kriesberg (eds), *Transformation and Peacebuilding: Moving from Violence to Sustainable Peace*, Routledge, New York, 2009; M. Ignatieff, 'The hate stops here', *The Globe and Mail*, 25 October 2001 <https://www.theglobeandmail.com/opinion/the-hate-stops-here/article763813/> (accessed 14 January 2018); M. Kaldor, *New and Old Wars: Organised Violence in a Global Era*, Polity Press, Oxford, 2001; J. Kaldor-Robinson, 'The virtual and the imaginary: the role of diasporic new media in the construction of a national identity during the break-up of Yugoslavia', *Oxford Development Studies*, 30(2), 2002, pp. 177–187; T. Lyons, 'Diasporas and homeland conflict', in M. Kahler and B. F. Walter (eds), *Globalization, Territoriality, and Conflict*, Cambridge University Press, Cambridge, 2006, pp. 111–132; T. Lyons. 'Conflict-generated diasporas and transnational politics in Ethiopia', *Conflict, Security, and Development*, 7(4), 2007, pp. 529–549; D. Newman. 'The resilience of territorial conflict in an era of globalization', in M. Kahler and B. F. Walter (eds), *Territoriality and Conflict in an Era of Globalization*, Cambridge University Press, Cambridge, 2006, pp. 85–110; Y. Shain, *Kinship and Diasporas in International Affairs*, University of Michigan Press, Ann Arbor, 2007; S. Turner. 'The waxing and waning of the political field in Burundi and its diaspora', *Ethnic and Racial Studies*, 31(4), 2008, pp. 742–765; A. Wennmann. 'The political economy of conflict financing: a comprehensive approach beyond natural resources', *Global Governance*, 13, 2007, pp. 427–444.
5. A. Smith, *The Ethnic Origins of Nations*, Blackwell, Oxford, 1986.

6. Lyons, 'Diaspora'.; Turner, op. cit.
7. H. Adelman, 'Why refugee warriors are threats', *The Journal of Conflict Studies*, 18(1), 1998, pp. 49–69.
8. Collier and Hoeffler, 'Greed'.
9. Turner, op. cit.
10. Ignatieff, op. cit.; Bock-Luna, op. cit.
11. J. Hall, 'Are migrants more extreme than locals after war? Evidence from a simultaneous survey of migrants in Sweden and locals in Bosnia', *Journal of Conflict Resolution*, 60(1), 2016, pp. 89–117.
12. M. Barslund, M. Busse, K. Lenaerts, L. Ludolph, and V. Renman., *Labour Market Integration of Refugees: A Comparative Survey of Bosnians in Five EU Countries*, CEPS Special Report No. 155, 9 December 2016.
13. It has been impossible to properly address this specific question in this paper due to the lack of enough country-specific data; there is the need for a much more ambitious—and better resourced—data collection process.
14. In certain cities outside of Serbia (London, Toronto, Washington, DC, Chicago, New York and San Francisco), diaspora members voted on 1 April 2017.
15. Serbian Government, MFA (Office for the collaboration with the diaspora and with Serbs in the Region)—The vote from abroad <http://www.dijaspora.gov.rs/lat/glasanje-iz-inostranstva/> (accessed 23 January 2018).
16. RIK (Serbian Electoral Committee)—Results of the 2017 Presidential Election <http://www.rik.parlament.gov.rs/izbori-za-predsednika-republike-2017-konacni-2004.php> (accessed 23 January 2018).
17. Media analysis agency Kliping found that an astonishing 51% of airtime dedicated to the 2017 presidential elections was dedicated to Aleksandar Vučić, and the figure increases to 67% when including his appearances as Prime Minister <http://www.danas.rs/izbori.1180.html?news_id=342219&title=Apsolutna%20dominacija%20Vu%C4%8Di%C4%87a%20na%20TV> (accessed 23 January 2018). The message conveyed by Vučić's campaign through a TV spot was that Serbia could not be governed by two pilots—implying that he would be keeping the helm in the future as well.
18. V. Ilić, 'Igrajmo se zanimljive izborne geografije u Srbiji' [Let's play an interesting election geography game in Serbia], *VICE Serbia*, 2 May 2017 <https://www.vice.com/rs/article/igrajmo-se-zanimljive-izborne-geografije-u-srbiji> (accessed 23 January 2018).
19. Blic, 'Janković: Potrebna nam je država, a ne bratija' [We need a State, and not informal networks], *Blic*, 15 February 2017 <https://www.blic.rs/vesti/politika/jankovic-potrebna-nam-je-drzava-a-ne-bratija/z7dk8e4> (accessed 23 January 2018).
20. Survey 1. *Serbian Diaspora—Presidential Elections 2017*. Online survey conducted by the author on 25 March–1 April 2017. Respondents were Serbian adult citizens living abroad. $N = 322$. Full data available in private archive.
21. Survey 2. *Serbian di-aspora 2017—Follow-up*. Follow-up online survey conducted by the author on 2–7 May 2017. Respondents were a random selection of respondents to Survey 1. $N = 83$. Full data available in private archive.
22. T. Prelec, 'Voters are split between two distinct groups: pro- and anti-Vučić', in 'Experts react: Aleksandar Vučić wins Serbia's presidential election', *LSE EUROPP—European Politics and Policy*, 3 April 2017 <http://blogs.lse.ac.uk/europpblog/2017/04/03/experts-react-aleksandar-vucic-wins-serbias-presidential-election/#Four> (accessed 23 January 2018).
23. Responses were furthermore collected from the following countries: Australia, Belgium, Bosnia and Herzegovina, Bulgaria, Croatia, Cyprus, Czech Republic, Denmark, Estonia, Faroe Islands, Finland, Greece, Hong Kong, Hungary, Iceland, Indonesia, Ireland, Israel, Luxembourg, Malta, Mexico, Monaco, Montenegro, Poland, Portugal, Qatar, Romania, Russia, Singapore, Slovakia, Slovenia, South Korea, Spain, Thailand, and Turkey.
24. The quotes are selected for their generally typical and representative nature.

25. The Editorial Board, 'A Serbian Election Erodes Democracy', *The New York Times*, 9 April 2017 <https://www.nytimes.com/2017/04/09/opinion/a-serbian-election-erodes-democracy.html> (accessed 23 January 2018).
26. Bock-Luna, op. cit.
27. See this graph by twitter user Filip Perisic: 'How to fight for your voting rights if you live abroad' <https://twitter.com/OnlineTulip/status/837800761983664130> (accessed 23 January 2018).
28. Glasajmo Iz Inostranstva—Facebook page: <https://www.facebook.com/GlasajmoIzInostranstva/>; Twitter page: <https://twitter.com/glasajmo> (accessed 23 January 2018).
29. Email interview with the author, 23 March 2017. Private archive.
30. Dijaspora Glasa—Facebook page: <https://www.facebook.com/dijasporaglasa/?ref=br_rs> (accessed 23 January 2018).
31. T. Prelec, 'The Macedonian surveillance scandal that brought down a government', *Computer Weekly*, June 2017 <http://www.computerweekly.com/feature/The-Macedonian-surveillance-scandal-that-brought-down-a-government> (accessed 25 April 2018).
32. Expert speaker at the roundtable on 'The UK and the Future of the Western Balkans', International Relations Committee, UK House of Lords, 17 January 2018 (event held under Chatham house rules). This issue—how attitudes towards emigration have changed among Macedonian citizens in the aftermath of the government's change—is a clear potential avenue for future research.
33. Ibid.
34. R. Crescenzi, N. Holman and E. Orru, 'Why do they return? Beyond the economic drivers of graduate return migration', *The Annals of Regional Science*, 59, 2017, p. 603.
35. These issues resonate with the problems encountered in recent times by several diasporas from South-East Europe, e.g. by Romanians and Bulgarians (in the latter case, a campaign by members of the diaspora has resulted in a partial improvement of the voting regulations): it is clear that this is not only a Serbian problem, and a comparative analysis would be extremely welcome.
36. A. Efendic, B. Babic and A. Rebmann, *Social Capital, Migration, Ethnic Diversity and Economic Performance: A Multidisciplinary Evidence from South Eastern Europe*, Peter Lang AG International Academic Publishers, Oxford, 2017.
37. S. Carter, 'The geopolitics of diaspora', *Area*, 37(1), 2005, pp. 54–63.
38. However, further research needs to be done on the views of the Croatian diaspora as well, considering the very high number of young Croats who have left the country to work in other EU countries in the last years. It is highly likely that the composition of the diaspora has changed, and their political views with it.
39. The movement *Ne Davimo Beograd* have stated that approximately 75% of their activists have either had an experience abroad or are currently enrolled in a study programme abroad. Interview by the author with coordinators of the 'Ne Davimo Beograd' movement. Belgrade, 12 April 2017, and follow-up via email.

Acknowledgements

The author would like to thank the many Serbian diaspora groups and individuals that have enthusiastically contributed to the dissemination of the survey, as well as the LSE's EUROPP-European Politics and Policy Blog (edited by Dr Stuart Brown) for providing an early avenue for discussion.

Disclosure statement

No potential conflict of interest was reported by the author.

Funding

This work was supported by the UK's Economic and Social Research Council (ESRC) [grant number ES/J500173/1].

The Effects of Remittances on Education in a Post-Conflict Society: Evidence from Bosnia-Herzegovina

Nermin Oruc, Ian Jackson and Geoffrey Pugh

ABSTRACT
This paper analyses the effects of remittances on the educational enrolment of children in Bosnia-Herzegovina, where a process of forced migration made the relocation decision exogenous. Accordingly, this study has no need of methods to address the endogeneity of remittances. Hence, the approach taken means that the measure of educational enrolment is regressed on a set of individual- and household-level variables. In addition, a new approach to estimation is introduced, whereby the effect of remittances is calculated for each income quintile. It is found that the relationship between remittances and educational enrolment is strong among households from the fourth quintile, which includes households just outside the risk of poverty, while for those in poorer quintiles the effect is not significant.

Introduction

This study investigates the effects on human capital formation of remittances arising from *forced* migration. The initial original contribution is the research of forced migration as a field of enquiry, which is largely neglected within the economics of migration. Secondly, it extends the literature analysing migration effects on human capital formation in the home country. Previous studies have had to grapple with the potential *endogeneity* of households' migration decisions to decisions regarding the education of their children.[1] In contrast, in this study, the migration decision is a brutal and wholly *exogenous* imposition on households. Accordingly, this study approaches the human capital consequences of migration free of the difficulties confronting the estimation of instrumental variables.[2] In this sense, it offers an identification strategy akin to a natural experiment.

The context of this study is forced migration arising from ethnic cleansing in Bosnia and Herzegovina (BiH) in the mid-1990s. Emigration from the Balkans has a long history[3] and the region has traditionally been a source of migrants to Western Europe and the United States since the late nineteenth century. The temporary migration of labour intensified in the early 1960s, during the implementation of the 'Gastarbeiter' (guest-worker) programme created by the governments of West Germany and

Yugoslavia, offering temporary jobs in Germany to Yugoslav workers. Most recently, the dissolution of the former Yugoslavia and subsequent war in Bosnia-Herzegovina forced around 50% of the total BiH population into migration. As a direct consequence, BiH is ranked as one of the leading countries in the world in terms of the size of its diaspora compared to the home country population (between 40% and 50%, depending on the source), as well as in terms of remittance inflows as a share of gross domestic product (GDP) (11%). In the mid-1990s, forced migrants, defined as the need to relocate due to violence and/or the threat of violence, came to account for around 75% of the Bosnian diaspora and an even greater share of remittances. This has subsequently made up a large proportion of total income for many households in the country. However, relatively little is known about the use and effect of remittances at the household level. This paper provides specific empirical evidence of the effect of remittances on the school enrolment of children.

As remittances are becoming the most important source of foreign capital for developing countries, outstripping the scale of overseas development aid and foreign direct investment according to the official data of the BiH Central Bank, then the issue of their effects at both the microeconomic and the macroeconomic level has received copious attention from researchers in the area of migration studies. The literature on remittances has recently changed from its traditional focus on the motivation for remitting to analysis of the use of remittances. There are numerous studies that analyse the effect of remittances on different aspects of a remittance-receiving economy, such as Funkhouser's study of the effect of remittances on the labour supply of remittance-receiving households in Nicaragua,[4] or Chami *et al.* on the effect of remittances on growth and productivity.[5]

Besides the research mentioned above, an increasing number of studies analyse the effect of remittances on investments in human capital formation. The first empirical studies on the relationship between remittances and educational attainment simply related remittance receipts to the educational attainment of children and assigned any effect found to the remittances.[6] These studies argued that remittances positively influence households' educational investment decisions through the relaxation of the liquidity constraint,[7] and reducing children's labour participation.[8] In this context, remittances, as an additional source of household income, help to provide additional disposable income. As a result, this induces investment in children's education leading to a positive relationship between remittances and children's educational attainment and a negative relationship between remittances and child labour supply. This result is known as the 'income effect'. Nevertheless, if household budget constraints are not binding, remittances should not have any significant effect on education expenditures. Consequently, if remittances are received by both households that face a budget constraint and households not having such a constraint, as potentially in the case of BiH, then an empirical specification that does adequately distinguish between these two types of household will likely yield an insignificant effect of remittances on education. In addition, more recent studies attempt to control for parental absence effects, besides the effect of relaxation of liquidity constraints.[9] This specific effect of remittances, which arises from the effect of a parent's absence, is termed the 'disruption effect'.[10]

The evidence on the effect of remittances on the educational attainment of children is mixed, and depends on the approach taken by researchers to the analysis. Several studies found a positive relationship between remittance receipts and educational enrolment.[11] Alternatively, some studies found no significant effect of remittances on education, while accounting for the effect of migration of parents. Moreover, several studies controlled for the gender of a child and found that the effect is more significant for female children,[12] while other studies found no significant difference between the genders of pupils.[13] This focused review of the literature suggests that the results are sensitive to the model specification (i.e. whether the model controls for the type of the household according to its budget constraint), the data used (i.e. whether migration is an endogenous decision or not), and the country analysed. This paper provides the first empirical evidence for BiH, addressing the issues of model specification (using income quintiles to distinguish between budget-constrained households and those that are not constrained) and of the data used (by using data about forced migrants, where the migration decision is exogenously determined).

Therefore, this paper makes three important contributions to the empirical study of the relationship between migration and human capital formation in sending countries. First, it investigates the effects of remittances on human capital formation in a country affected by *forced* migration, which allows us to assume *exogeneity* of the migration decision. Second, it introduces a *new model specification*, whereby interactions between the receipt of remittances and income quintiles are used to identify the effect of liquidity constraints, as theoretically described. Finally, it uses equivalence scales for measuring *per child* household consumption of education more precisely.

The paper is structured as follows. The next section describes briefly the scenarios of education and remittances in Bosnia-Herzegovina in order to provide necessary background for the study. The subsequent section presents a short theoretical discussion of the relationship between remittances and education. Then we describe our empirical model and estimation strategy, followed by the results of the empirical analysis. The final section presents the conclusions.

Background: remittances and education in Bosnia-Herzegovina

Bosnia-Herzegovina is ranked 23rd in the world in terms of receiving remittances as a share of GDP. The remittances amount to more than €2 billion annually, which is around 11% of Bosnian GDP.[14] Moreover, remittances represent the most significant inflow to BiH, as they were six times larger than foreign direct investment and three times larger than development assistance to this country in 2017. These remittance inflows are a significant source of income for a large proportion of the BiH population. The data from the 2004 round of the Living in BiH survey (LBiH) were used to analyse the effect of remittances on poverty and inequality, showing that approximately one-third of households in Bosnia-Herzegovina receive remittances.[15] The average value of remittances received is about 100 KM per month.[16]

The figures presented in Table 1, in the absence of more comprehensive studies on the relationship between remittances and poverty in Bosnia-Herzegovina, were produced by using the LBiH dataset in order to provide background information for the study of the relationship between remittances and educational attainment completed in this paper. Accordingly, the remittances sent to Bosnia-Herzegovina are not pro-poor. A larger proportion of non-poor households receive remittances. Moreover, the average

Table 1. Share of households receiving remittances by category (*N* = 7702).

Criteria	Recipients (% of the sample)	Average amount received per month (KM)
All	33.5	100.54
Poor	28.3	52.98
Non-poor	34.3	105.98
Head of household		
• Primary education	32.3	83.27
• Secondary education	30.5	105.93
• Tertiary education	26.0	104.92
Rural	36.4	112.24
Other urban	32.4	86.85
Capital (Sarajevo)	31.9	105.22
Female-headed households	44.7	98.39
Male-headed households	29.6	101.69

Note: Poor households are defined as those whose annual household income was below the relative poverty line (60% of median income). Non-poor households are those with an income above the relative poverty line.
Source: Authors' calculations based on the LBiH 2004 survey.

amount received is twice as large as the amount received by poor households. Regardless, a larger proportion of households with a less well-educated head receive remittances, albeit a slightly lower average amount compared to those with a more highly educated head. Also, a significantly larger number of female-headed households receive remittances than do male-headed households.[17] Households in rural areas receive more remittances, both in terms of the proportion of households that receive them and in terms of the average amount, than do households in urban areas. These outcomes tend to suggest that remittances, although not contributing to poverty reduction overall, may be expected to have a positive effect on reducing differences in poverty incidence between rural and urban areas.

Table 2 shows that remittances do not contribute to a decrease in inequality, as the average amount of remittances tends to increase from the poorest to the richest decile. Also, the proportion of households receiving remittances in each decile, with the exception of the richest quintile, generally increases as we move from poorer to richer deciles. This information is important for the argument that remittances influence educational investments only of those households that fall into a specific income bracket, which will be explained in detail later.

To better understand the context of the study, we also briefly describe the educational system in BiH. The main characteristics of the educational system in Bosnia-Herzegovina as a result of the war are a reluctance to reform the educational system inherited from former Yugoslavia and the burden that the political organization of the country puts on the educational system.[18] According to the Dayton Peace Agreement, Bosnia-Herzegovina is a country organized into two entities, the Federation of BiH (FBiH) and the Republika Srpska (RS), where the Federation of BiH is comprised of ten cantons. Education in RS is organized and controlled at the entity level, whereas in

Table 2. Recipients of remittances by income decile.

	Decile										
Criteria	1	2	3	4	5	6	7	8	9	10	Average
Recipients (% of the sample)	27.4	33.6	30.8	35.2	37.6	36.6	38.2	34.0	35.4	26.3	33.5
Average amount received (KM)	51.7	74.6	62.6	103.5	83.0	106.7	107.3	107.2	140.5	169.6	100.5

Source: Authors' calculations based on the LBiH 2004 survey.

Table 3. Official statistics about educational enrolment in BiH, 2015.

Education level	Number of schools	Number of students	Number of teachers	Student/teacher ratio
Primary	1,812	296,842	24,118	12.31
Secondary	311	144,231	12,773	11.29
Tertiary		108,008	9,581	11.27

Source: BiH Agency of Statistics: Education Statistics, 2015.

FBiH most education issues are regulated at the canton level. The lack of coordination between cantons and entities make any attempt at meaningful reform tremendously difficult.

The legacy of the incumbent Yugoslav system, with highly subsidized education at all levels, but without appropriate reforms in terms of adjustments to the new structure of the economy and labour market needs, along with the lack of coordination between different levels of authority, has resulted in significant mismatch between the skills produced by the education system and those required by the labour market. The phenomena of 'educated unemployment' and 'overeducated' individuals, as reported in the Labour Market Surveys, seem to be widespread.[19]

In a country without a national census since 1991,[20] it is extremely difficult to obtain relevant data on the literacy rates and educational structure of the population. According to the Census in 1991, about 10% of the population were illiterate. Small-scale household surveys conducted in the meantime suggest that this rate has declined to about 5%. Also, these surveys suggest that about 9% of the population are without education, more than 30% have primary education only, about 50% have secondary education and 8% have higher education. The net enrolment rate in primary school is about 97%, in secondary school about 75% and in higher education about 20%. Around 55% of students enrolled in secondary schools graduate on time. Official statistics about the number of schools, students and teachers in BiH for 2015 are presented in Table 3.

Child labour is relatively uncommon in Bosnia-Herzegovina,[21] so the drop-out rates are generally very low for children of primary school age (6–15 years old), compared to some other countries in which empirical evidence on the remittances effect has been investigated, notably in Latin America. As a result, the focus of this paper is on the educational attainment of young people aged 16 to 24. This age span encompasses young people in secondary and tertiary education where alternatives such as labour market participation are present and where drop-out rates are much more significant compared to primary education. The next section presents the empirical investigation.

Empirical analysis

Data

Our analysis of the effect of migration on the educational attainment of children in Bosnia-Herzegovina was conducted by using the dataset from the LBiH survey, conducted by the BiH Office of Statistics in 2004. The original sample of this survey was 7702 households, or 22,483 individuals, including 6169 individuals younger than 25. The number of education-age (6–24) individuals was 5136, whereas the number of individuals aged between 16 and 24 was 2758. Additionally, this sample was reduced by excluding observations missing information about the child's enrolment. Hence, the final dataset used for the model estimation contains 1367 observations. The list of variables from the dataset used in the empirical analysis of the relationship between remittances and education, together with corresponding descriptive statistics, is presented in Table 4. The names of variables prefixed with the letter *c* indicate child-level information, while those with *h* indicate household-level information.

The proportion of households with a child in post-primary education that reported receipt of remittances was 9.2%. This share is in line with other reports about the share of households receiving remittances in BiH.[22] Notwithstanding, what is usually being under-reported is the amount of remittances, as suggested in previous studies. The average amount of remittances received by a household per is KM 121.78.

The model and estimation strategy

Equation 1 is a probit regression estimating the determinants of enrolment.[23]

$$\begin{aligned} yy_i = \beta_0 + \beta_1 \text{hrem} + \beta_2 \sum_{(i=16)}^{24} \text{cage} + \beta_3 \text{female} + \beta_4 \text{houseown} + \beta_5 \text{phoneown} \\ + \beta_6 \text{hhprim} + \beta_7 \text{hhsec} + \beta_8 \text{hhtert} + \beta_9 \text{hhage} \quad + \beta_1 0\text{hhemp} + \ \beta_{11} \text{hhsize} \\ + \beta_{12} \text{capital} + \beta_{13} \sum_{(i=1)}^{4} \text{quintiles} + \beta_{14} \sum_{(i=1)}^{4} \text{hrem* quintiles} + u_i \qquad (1) \end{aligned}$$

The dependent variable in Equation (1) is dichotomous, indicating whether a child is currently enrolled in a school (*cenr*). The data used for the estimation of the model arise from exogenous selection into migration, which avoids the potential problem of an endogenous relationship between remittances and education.[24] A pupil's age (*cage*) and

Table 4. Summary descriptive statistics of variables.

Variable description	Variable name	*N*	Mean	Std Dev.
= 1 if child currently enrolled in education	cenr	1293	0.48	
Monthly amount of household's education expenditure per child, in KM	heepcx	1367	484.80	2151.988
Age of child	i.cage	1367	19.76	2.855
= 1 if child is female	cfemale	1281	0.48	
= 1 if an individual is of tertiary school age	ctedu	1367	0.16	
= 1 if a child lives in the capital city	capital	1303	0.21	0.408
= 1 if female-headed household	hfh	1367	0.01	
Household head's age	hhage	1272	47.89	9.898
Employment status of a household's head, 1 if employed	hhemp	1367	0.16	
= 1 if a household has a phone	hhphone	1367	19.76	2.855
= 1 if a household owns a house	hhouse	1303	0.82	
= 1 if a household head has primary education	hhprim	1367	0.21	
= 1 if a household head has secondary education	hhsec	1367	0.47	
= 1 if a household head has tertiary education	hhtert	1367	0.10	
Number of household members within a household	hhsize	1303	4.25	1.414
Household's monthly income per capita, in KM	hincpcx	1367	2389.66	2906.268
Number of children in the household	hnoc	1367	2.11	1.016
Number of school age individuals in the household	hnoc624	1367	0.15	0.447
Number of secondary or tertiary school age individuals in the household	hnoc1518	1367	1.67	0.723
= 1 if a household receives remittances	hrem	1293	0.09	
Monthly amount of remittances received by a household (in KM)	hrema	1367	121.78	632.929

Source: Own calculations.

quintiles were entered as individual categories, with age 24 and the first quintile as benchmark categories, respectively. A pupil's gender (*female*) is also included in the model, as a standard variable that controls for differences in enrolment between boys and girls.

The main variable of interest in this model is receipt of remittances by a household. Given that migration from BiH was predominantly a migration of entire families as refugees, the receipt of remittances can be considered as an exogenous income gain for households and captured by the remittances receipt variable.[25] There are three alternative variables for the receipt of remittances: first is a dummy variable for receipt of remittances ($hrem_i$); second is a continuous variable for the average monthly amount of remittances received in a year ($hrema_i$); third is a variable for the amount of remittances per month relative to a household's budget ($hremarel_i$). Since the initial analysis revealed that the continuous variables (*hrema* and *hremarel*) did not appear to be statistically significant, the dummy variable *hrem* is used in further analysis. The use of the dummy variable instead of the amount is justified by the experience reported in previous studies, which tends to suggest that remittance-receiving households are likely to underreport the amount of remittances received.[26]

The main argument of the 'brain gain' literature regarding the positive effect of remittance receipts on educational investment is that remittances have a positive effect via relaxation of liquidity constraints faced by households. However, this effect should not be expected for all households receiving remittances. The liquidity constraint is binding for low-income households only. Conversely, households at higher levels of income distribution may not face a liquidity constraint, in which case the receipt of remittances is not expected to change their education investment decisions. In addition, any receipt of remittances by households at the bottom of the income distribution might not change their education investment behaviour, as the amount of remittances received may not be sufficient to allow such investment. Remittances sent to such households are usually purely altruistically driven and are spent on basic needs and current consumption.[27] Therefore, the receipt of remittances is expected to have an impact on investments in education only among households that are within a limited range of the income distribution—those where the pre-remittance level of income is sufficient for current consumption and the receipt of remittances is thus an addition to income that enables investments in education. Accordingly, the empirical model to be estimated includes a list of dummy variables for income distribution quintiles (Σq_i), as well as a set of interactions between the receipt of remittances and each consumption quintile ($\Sigma q_i hrem$). Specification of our model with these variables enables us to capture the difference in the effect of remittances between households at different levels of the income distribution.[28] The strongest effect of remittances on education is expected for households that are in a specific quintile, where they are not rich (and still face liquidity constraints), but are not so poor as to be forced to spend all the remittances received on basic needs. This model specification is another original contribution of this paper.

We also included a set of control variables about the household's and household head's characteristics that may affect a child's school enrolment. The household head's educational level ($hhedu_i$) is considered as a proxy for household wealth. Parents with higher education are expected to have a higher probability of wage employment and to earn higher wages than less well-educated parents. Consequently, households with more highly educated parents may be less likely to face liquidity constraints and therefore are more likely to invest in their children's' education. As the majority of studies investigating poverty incidence and inequality between male- and female-headed households in BiH[29] suggest a positive relationship between being a female-headed household and household income, a dummy variable for the gender of the household head (hfh_i) is included. The household size ($hhsize_i$) variable is expected to have a negative effect on investment in children's education. The larger the household, the lower is the household budget per capita and thus the stronger is the liquidity constraint, all else remaining equal. Furthermore, the number of school-age children ($hnoc624_i$) also affects educational investments as the reservation income for education is lower per capita in households having more school-age children. A household's employment status ($hhemp_i$) is also included. The type of area ($capital_i$) is considered to capture both the effect of cultural and infrastructural differences and the effect of economic differences. Since previous studies[30] have shown that poverty is present even more in non-capital urban areas than in rural areas, the actual welfare gap in BiH is between people living in the capital city and those living in other areas. All else remaining equal, households in rural areas are expected to invest less in the education of children.

Results and discussion

The results of probit estimation of Equation (1) are reported in full in Table 5. First, we comment briefly on our control variables. Then, we discuss the impact of remittances on school enrolment in more detail.

The main demographic factors influencing enrolment are the age of the child, gender and the economic situation of the household. Compared to those aged 24 (the omitted category), children of secondary school age (16–18) are much more likely to be enrolled, while others of tertiary age (19–23) are equally likely to be enrolled. The positive sign of the gender dummy (*female*) in the estimates from the survey suggests that girls are more likely to be enrolled in school, *ceteris paribus*, than are boys. This is consistent with the fact that females from the sample (age 15–24) face much less favourable labour market conditions,[31] thus employment opportunities have less of a diverting influence on their enrolment choice. The positive effects of parental education after primary level (estimated by the two variables *hhsec* and *hhtert*) are also in line with theoretical predictions and previous studies. The negative effect of a household's size on pupil enrolment may capture the effect of intra-household allocation of income, where a child's education competes with other needs. In addition, residence in the capital city and owning a phone are also statistically significant and positively associated with enrolment. The apparently perverse effect of house ownership on enrolment may reflect the predominance of house owners in the sample (82%), and—with the exception of quintile 5 (97%)—their relatively equal distribution across income quintiles (Q1 = 78%, Q2 = 72%, Q3 = 76%, Q4 = 86%).

The estimated coefficients from the probit regression do not have any intuitively appealing quantitative interpretation. For the control variables discussed above, it is sufficient to comment on their qualitative effect. Conversely, for our variables of interest we need to provide some quantitative interpretation. In other words, we need to know

Table 5. Table with results of model estimations.

Dependent variable is a dummy variable taking value of 1 if child is currently enrolled in education	Name	Coefficient
= 1 if a household receives remittances	hrem	0.154
Age of child (24 years is benchmark)	i.cage	−1.311***
Age 16		2.061***
Age 17		2.011***
Age 18		0.859***
Age 19		0.165
Age 20		0.155
Age 21		−0.232
Age 22		−0.106
Age 23		−0.001
= 1 if child is female	cfemale	0.362***
= 1 if household owns a house	houseown	−0.294**
= 1 if household owns a phone	own_phone	0.322**
= 1 if household head has primary education	hhprim	0.281*
= 1 if household head has secondary education	hhsec	0.704***
= 1 if household head has tertiary education	hhtert	1.126***
Age of household head	hhage	0.025***
= 1 if household head is employee	hemp	−1.455***
Household size	hhsize	−0.247***
= 1 if household lives in the capital city	capital	0.463***
= 1 if household head is married	hhmarried	−0.142
Quintile 2	q2	−0.045
Quintile 3	q3	0.032
Quintile 4	q4	0.605***
Quintile 5	q5	0.550***
Quintile 1 x hrem	q1xhrem	
Quintile 2 x hrem	q2xhrem	−0.454
Quintile 3 x hrem	q3xhrem	−0.163
Quintile 4 x hrem	q4xhrem	0.687
Quintile 5 x hrem	q5xhrem	1.227
Number of observations		1,070
Pseudo R^2		0.412
Pearson chi2(1130)		609.86
Prob > chi^2		0.000
Correctly classified (%)		82.06

Note: statistically significant at: ***1% level, **5% level, and *10% level of significance.
Source: Own calculations.

not just whether our estimates are statistically significant and of one or other sign, but also whether they reveal practically and logically significant effects. To this end, for households in each income quintile we use the regression results to derive the average marginal effects of receipt of remittances on the probability of childens' enrolment in education. For households in each income quintile, after taking into account the average values of all the covariates, we compare the average probabilities of enrolment for households in each income quintile (i) receiving remittances and (ii) not receiving remittances. [32]Figure 1 reports these estimated comparisons: for households in each income quintile, the respective dot depicts the average marginal effect of remittances on the probability of school enrolment; and the vertical bars depict the associated confidence intervals (such that the shorter the bar the more precise the estimate, and estimates with bars overlapping the zero reference line indicate an estimate that cannot be statistically distinguished from zero). These comparisons are given in terms of the probabilities of enrolment (e.g. an estimate of +0.2 is to be interpreted as a remittances effect increasing the probability of enrolment by 20%).remittances.

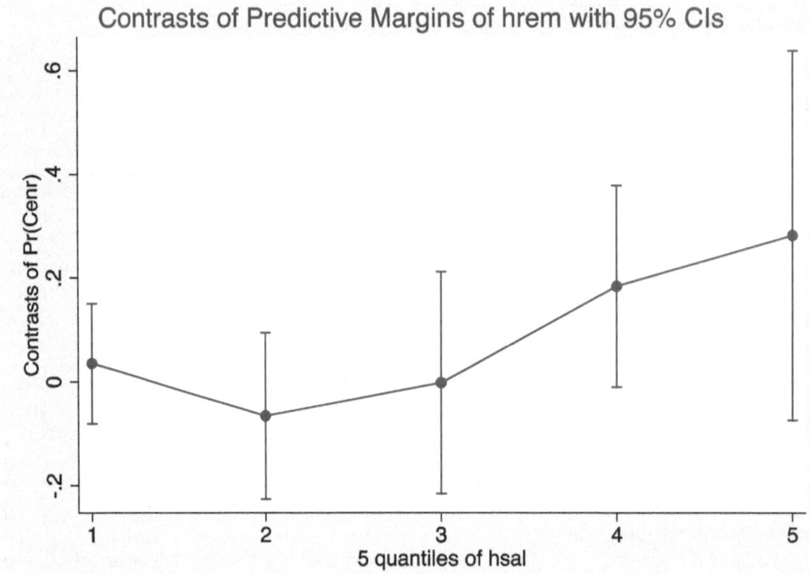

Figure 1. Difference in the marginal effects of receiving remittances compared to marginal effects of not receiving remittances.

Figure 1 shows that, as we hypothesized, at the lower income quintiles (1, 2 and 3) there is no enrolment effect of remittances; not only are the estimated differences tiny, but each of them lacks statistical significance (in each case, confidence intervals are almost symmetrical around zero). In contrast, households receiving remittances in income quintile 4 are 18% more likely to enrol their children in education ($p = 0.063$, i.e. borderline at the conventional 5% level of significance). Likewise, consistent with our hypothesized effects, the evidence for a remittance effect in the highest income quintile is statistically weaker ($p = 0.120$, i.e. somewhat beyond the conventionally acceptable threshold for statistical significance) which, if interpreted strictly, indicates that the effect is not significantly different from zero.

Of course, the statistical significance of econometric estimates is to be interpreted in the light of sample size. The numbers of households receiving remittances in income quintiles 1, 2, 3 and 4 are, respectively, 48, 28, 18 and 20. Hence, the lack of statistical significance of the tiny remittance effects for quintiles 1, 2 and 3 compared to the significance level of the large quintile 4 effect is not an artefact of the different numbers of observations supporting the estimates. Accordingly, we argue that the contrast between our estimated effects for quintiles 1, 2 and 3 and our estimated effects for quintile 4 is valid. Conversely, our sample includes only five households receiving remittances in income quintile 5. This makes it likely that the relatively imprecise estimate of the remittance effect on school enrolment for the highest income quintile reflects a very limited sample size. We conclude that our hypothesized variation of remittance effects on enrolment by income receives strong support from the comparison between the zero effects for quintiles 1, 2 and 3 and the large effect for quintile 4, but less robust support from a comparison of the quintile 4 and quintile 5 effects.

These outcomes are broadly consistent with the presence of differential liquidity constraints at different levels of income, as explained above. Statistical insignificance and the negligible size of the marginal effects of receiving remittances compared to the marginal effects of not receiving remittances in the first to third income quintiles suggest that remittances are not sufficiently strong to remove liquidity constraints from households with a below average income. Moreover, statistical significance of the large difference in the marginal effect of receiving remittances compared to the marginal effect of not receiving remittances in the fourth quintile shows that, for relatively richer households, remittances make a substantial difference in a child's probability of enrolment. However, for the highest income households in quintile 5, we find weak evidence that liquidity constraints may not be a binding influence on school enrolment, so remittances make no difference.

Conclusions

This paper presents empirical evidence regarding the effect of remittance receipts on enrolment in secondary and tertiary education in Bosnia-Herzegovina. The approach provides an empirical contribution to the literature by using data in which the migration decision is exogenous, consequently producing empirical results by estimating a model free of the difficulties confronting instrumental variables estimation. Moreover, the model specification introduces a new way of localizing the liquidity constraint effect by income quintiles.

The findings presented in this paper suggest that remittances have a significant effect on educational attainment. However, the positive effect of remittances is limited to a specific group of households: namely, relatively well-off households facing a liquidity constraint that can be relaxed by remittances to an extent sufficient to increase substantially the probability of school enrolment. Poorer households cannot afford education even when they receive remittances, as they use these receipts for covering basic needs, while the richest households can afford education even if they do not receive remittances. The model specification used in this paper, where quintiles of income distribution were used to capture the effects of remittances on educational enrolment at different levels of income, is more consistent with the theoretical predictions about the remittance effects than are the specifications used in previous empirical studies.[33] Model misspecification by not including income distribution (quintiles) might be a reason why some previous studies did not find any influence of remittances on the education of children.

Notes

1. More discussion on the issue of potential endogeneity in studies of migration is provided in the empirical analysis section later in the paper. For now, we just note that endogeneity refers to the possibility that, for example, the dependent variable (enrolment) and the independent variable of interest (remittances) may simultaneously determine each other (i.e. because households make simultaneous and mutually conditioning decisions regarding their children's education and decisions to migrate resulting in remittances). Ignoring such a relationship between variables in the model may result in biased estimates. However, in our case, migration is forced onto the

household, which means that migration and its consequences (including remittances) can be regarded as exogenous to the household rather than as an endogenous outcome of household decision making.

2. Previous studies on the effect of remittances on education have long dealt with the endogeneity of the migration decision as one of the main issues, although instruments widely used in earlier studies were later found to be invalid. For an example of earlier studies see the historical migration account by G. Hanson and D. Woodruff, *Emigration and Educational Attainment in Mexico*, National Bureau of Economic Research Working Paper, Cambridge, MA, 2003. For later corrective studies see C. Calero, A. S. Bedi and R. Sparrow, *Remittances, Liquidity Constraints and Human Capital Investments in Ecuador*, IZA Discussion Paper 3358, Bonn, 2008; and the discussion in M. R. V. Borromeo, *Remittances and the Educational Attainment of Children in the Philippines*, SLU Uppsala, Department of Economics, Uppsala, 2012.
3. See, for example B. Repe,' Migrations in the territory of former Yugoslavia from 1945 until present time, European Seminar of Teachers "Being a migrant in Europe", 6-9 March 2002, Faculty of Arts and Sciences', Séminaire européen d'enseignants 'Etre migrant(e) en Europe', 6–9 March 2002, Faculté des Lettres et Sciences Humaines, Université de Neuchâtel, Neuchâtel.
4. E. Funkhouser, 'The effect of emigration on the labor market outcomes of sender households: a longitudinal approach using data from Nicaragua', *Well-Being and Social Policy*, 2(2), 2006. pp. 5–25.
5. R. Chami, C. Fullenkamp and S. Jahjah, *Are Immigrant Remittance Flows a Source of Capital for Development?* IMF Staff Papers 52(1), Washington, DC.
6. See, for instance, A. Cox-Edwards and M. Ureta, *International Migration, Remittances, and Schooling*, NBER Working Papers 9766, 2003.
7. 'Liquidity constraint' in the context of remittances should be interpreted as the inability of households to borrow and to invest in productive activities, including education, due to a lack of financial resources and corresponding lack of ability to collateralize loans.
8. See, for example: Hanson and Woodruff, op. cit.; Cox-Edwards and Ureta, op. cit.; also D. McKenzie and H. Rapoport, *Can Migration Reduce Educational Attainments? Depressing Evidence from Mexico*, Stanford Center for International Development, Working Paper 274, 2006; C. Amuedo-Dorantes, A. Georges and S. Pozo, *Migration, Remittances and Children's Schooling in Haiti*, IZA Discussion Paper 3657, 2008.
9. For instance, C. Amuedo-Dorantes and S. Pozo, *International Migration, Remittances and the Education of Children: The Dominican Case*, Working Paper, Western Michigan University, 2006.
10. Y. Lu and D. J. Treiman, *The Effect of Labor Migration and Remittances on Children's Education among Blacks in South Africa*, California Center for Population Research Online Working Paper CCPR-001-07, 2007.
11. As examples Cox-Edwards and Ureta, op. cit.; Calero *et al.*, op. cit.
12. For instance, P. Acosta, *Labor Supply, School Attendance and Remittances from International Migration: The Case of El Salvador*, World Bank Policy Research Working Paper 3903, Washington DC, 2006.
13. See A. Cuecuecha, *The Effect of Remittances and Migration on Human Capital: Evidence from Mexico*, ITAM Working Paper, 2008.
14. Annual inflow of international remittances in 2016, according to the BIH Central Bank, was KM 2.4 billion, while the World Bank estimates KM 3.3 billion. On the other hand, the World Network of Bosnian Diaspora estimates the total net inflow as at least 6 billion. Divide by 2 to convert to euros.
15. There are more recent rounds of the BiH Household Budget Survey (in 2007, 2011 and 2015). However, the share of remittance-receivinghouseholds with post-primary age children in these datasets is below 5%, while in 2004 the share is 10%. For that reason we decided to use the 2004 data.

16. The BiH currency, KM, is a 'convertible mark', with a fixed exchange rate of approximately 2 KM to one euro.
17. We might hypothesize this as being due to two circumstances: female-headed households have husbands working abroad who send remittances to support their families back home; or single-person widow households supported by adult children working abroad. These mechanisms have been found to be widespread in the gendered remittance dynamics of Albania. See J. Vullnetari and R. King, *Remittances, Gender and Development: Albania's Society and Economy in Transition*, I. B. Tauris, London, 2011.
18. For example, EU reports on the progress in implementation of the Bologna process as well as other reports provide evidence on the lack of effort and commitment on the part of successive Bosnian governments to reform and improve the effectiveness of the country's educational system.
19. C. S. Fan and O. Stark, 'International migration and "educated unemployment"', *Journal of Development Economics*, 83(1), 2007, pp. 76–87.
20. The latest BiH Census was taken in 2013, but the data are not yet available.
21. UNICEF, Child Labour Statistics based on their Multiple Indicators Cluster Survey (MICS), accessed on 15 January 2018.
22. See N. Espinova, J. Ray and A. Pugliese, *Gallup World Poll: The Many Faces of Global Migration*, IOM Research Series, 43, 2011; I. Gedeshi and E. Jorgoni, *Social Impact of Emigration and Rural-Urban Migration in Central and Eastern Europe: Albania*, European Commission DG Employment, Social Affairs and Inclusion, Brussels.
23. Besides estimating the effect of remittances on the child's school enrolment, as is done in previous studies, we initially intended to go a step further and estimate the relationship between remittances receipt and educational investments by a household while controlling for selection bias by using Heckman's two-stage procedure. However, the number of observations for children from remittance-receiving households was not sufficient to estimate this model.
24. As indicated already, the endogeneity problem in research into the effects of remittances is well-known and has been addressed in different ways depending on the context of the particular research setting. For example, 'reverse causality' (or simultaneity) may arise when households strategically send a household member to work abroad in order to raise the finance needed for education of another household member through remittances. In this case, it is not only that remittance receipts affect education, but also that educational investment intentions can affect remittances. Alternatively, unobservable household characteristics such as ambition, ability, risk preferences and attitudes towards investing in the future can affect both remittances and educational enrolment.
25. To correctly assess the effect of relaxing liquidity constraints, it is not sufficient to include a variable on remittance receipt. In fact, it is necessary to take into account that, in households with an absent member, remittances are not an exogenous addition to income, but the alternative source of household income to the situation when the absent member would stay within a household and earn a wage. Accordingly, absence of a member in a household that does not receive remittances may be considered as an income loss. As a result, besides accounting for receipt of remittances, the distinction between households that do and those that do not have absent members is necessary. To this end, counterfactual income for households that have absent members, compared to households without migrants, should be estimated. However, although it may be very important in the case of voluntary migration, it is not necessarily important in the context of forced migration. The migration decision in the context of forced migration can be considered as exogenous, and not based on any household's income gain strategy. Moreover, there is evidence that many households in Bosnia-Herzegovina receive remittances from relatives, which thus may be considered as an exogenous source of income. It is also supported by the idea that conflict-induced migration differs from voluntary migration in the sense that it is predominantly a migration of

entire families, instead of one or two members. Also, this affects the amount of remittances sent, as interfamily links are expected to be weaker than intrafamily links.

26. F. Borraz, 'Assessing the impact of remittances on schooling: the Mexican experience', *Global Economy Journal*, 5(1), 2005, pp. 1–30.
27. D. Cox, B. Hansen and E. Jimenez, 'How responsive are private transfers to income? Evidence from a laissez-faire economy', *Journal of Public Economics*, 88(9), 2004, pp. 2193–2219.
28. The average amounts of remittances received by households at different levels of the income distribution in BiH are given in Table 3.
29. See S. Smajic-Ermacona, 'Poverty among female-headed households in Bosnia and Herzegovina: an empirical analysis', *South East European Journal of Economics and Business*, 2(1), 2007, pp. 69–88.
30. See UNDP BiH, Rural Development in Bosnia and Herzegovina: Myth and Reality - 2013, National Human Development Report, UNDP-BiH, Sarajevo, 2013. Available at: http://www.ba.undp.org/content/bosnia_and_herzegovina/en/home/library/nhdr/nhdr-2013-resource.html
31. The unemployment rate among the female workforce in Bosnia-Herzegovina is higher than it is for males.
32. We implemented the procedure as follows:

    (a) We estimated the probit regression reported in Table 5. All variables were entered in factor form, which is necessary to compute the marginal effects using Stata's *Margins* command. Our variables of interest—remittances and income quintiles—are specified both individually and in interaction.
    (b) We used Stata's post-estimation *Margins* command to obtain the remittances effect at each quintile, using the following syntax: margins hrem, at(quintile = (1(1)5)) contrast(effects)

    We used *Marginsplot* to obtain a graphical display that plots the difference in the probability of enrolment between households receiving remittances and households not receiving remittances for each income quintile.
33. Notably Cox-Edwards and Ureta, op. cit.; Amuedo-Dorantes and Pozo, op. cit.; Cuecuecha, op. cit.

Disclosure statement

No potential conflict of interest was reported by the authors.

The Greek Debt Crisis and Albanian Return Migration

Kalie Kerpaci and Martin Kuka

ABSTRACT

Existing research on the relatively recent phenomenon of Albanian migrants returning to Albania is limited and focuses on the reasons for return, the characteristics of the return migrants, and on employment in general. Based on in-depth qualitative interviews with 25 returnees in Tirana, the study explores primarily, but not exclusively, the impact of the ongoing Greek debt crisis on the decision of Albanian migrants to return to Albania. More specifically, it analyses the reasons that motivated them to leave Greece; the barriers they faced while trying to reintegrate into Albanian society, which has undergone vast changes during the time of their absence; the way these barriers influence their future intentions; and, finally, the role of the Albanian government in their reintegration.

Introduction

A century ago migration was thought to be a 'one-way trip'—a movement toward Europe, Asia and North America, without a homeward return.[1] In the United States, the belief that the migrants do not wish, or have no reason, to return to their country of origin, which has been called 'the myth of non-return', led the study of migration to focus mainly on issues of adaptation and/or assimilation in the receiving society.[2] Currently, return migration is a widely discussed topic among researchers and politicians because of its impact on both receiving and sending countries.[3] Migration researchers began to turn their attention to this dimension of international mobility in the late 1970s, emphasizing the reasons for return[4] and the impact on the economic development of the sending countries.[5] The experiences of return migrants in their home countries have yet to be researched in depth.

During almost three decades of migration, since the first mass exodus in the early 1990s, many Albanians have returned home, either voluntarily or involuntarily. Some have settled, others have migrated again, or moved back and forth in search of work opportunities,[6] without declaring their intentions to authorities. This mobility makes any effort to measure return migration very difficult. According to the Albanian Census of 2011, the number of Albanians who had returned from Greece since 2001 was 86,417. The most significant increase was observed from 2008 to 2011, years that coincide with the onset of the Greek crisis, during which the annual number of returned Albanians climbed from 6565 to 17,955.[7] A significant problem with this data is that the relevant question in the census questionnaire asked only whether the participants had lived in another country, without a follow-up question about their status in that country. Hence, in the case of those who returned from Greece, we cannot know if they were labour migrants, students at Greek universities, or even employees at the Albanian embassy. We simply know that they had lived in Greece. Moreover, other reports provide different data. Research in 2012 by the Albanian Centre for Competitiveness and International Trade (ACIT) estimated the number of Albanians returning from Greece due to the crisis at around 180,000, around one-fifth of their estimate of the total number of Albanian migrants in Greece,[8] while the research of the Albanian Institute of International Studies estimated that number to be around 15% of the approximately half-million Albanians in Greece (i.e. about 75,000).[9] The significant discrepancies between these reports confirm the difficulties of measuring Albanian return migration. However, all reports converge on the same conclusion—that the tendency to return to Albania has intensified due to the Greek crisis.

Methodology

The methodology we adopted in our research was that of interpretivism, which entailed an understanding of the subjective meaning of social action,[10] the way people perceive the reality part of what they are.[11] We aimed to understand the meaning that Albanian return migrants attached to their actions, such as their decisions to return, to set up a business in Albania, etc., and to the difficulties they faced while trying to reintegrate into Albanian society.

In this paper, we consider as returned migrants those people living in Tirana, the capital of Albania, for more than a year. This time period is thought to be sufficient for the returned migrants to reflect upon their current lives, by weighing both the positive and negative sides of their decision to return.

We are aware of the limitations of generalizing the findings by focusing only on Tirana as the place of our research. For a deeper understanding of Albanian return migration, field surveys should expand to other cities as well. Nevertheless, we chose this location because a significant number of returnees, although not originally from Tirana, decided to return there, and this is consistent with the general pattern of population mobility in Albania, either internal or return migration, i.e. a tendency to move to the capital. Tirana is the only city in Albania that has seen a marked increase in population since the lifting of mobility restrictions that followed the fall of communism in 1991. All other Albanian cities, with the exception of Durrës,[12] have seen a significant decrease in population due to internal and external migration.[13] Moreover, employment and educational opportunities in Tirana are better compared to other cities in Albania. It is one of the most economically preferable areas to settle in, followed by the coastal districts, where returnees can be employed or self-employed in tourism or agriculture, and the rural south, which has also experienced relatively significant economic development.[14] Also, Tirana is transformed into a modern city, with a way of life similar to that in Greece. In this context, returned migrants prefer to settle in Tirana even though some of them came originally from other cities of the country.

We conducted 25 in-depth, semi-structured interviews with Albanian migrants who had returned from Greece, were settled in Tirana, and belong to the first generation of Albanian migrants. Only one of the returnees belonged to the so-called 'zero generation'. He and his wife followed their son to Greece and later back to Albania, in order to support him in his plans.[15] To select participants we used the snowball technique. Our sample consisted of 14 males and 11 females, most of whom were of working age. The interviews were conducted at places chosen by them, and were undertaken by the authors from August 2013 to September 2014.

Initially, we explained the goal of our research and assured the returnees about their anonymity. The language of the interview was chosen by them. They all preferred Albanian. Each interview lasted from one to two hours, and was divided into two parts: the first focused on their living experiences in Greece, and the second focused on the process of their reintegration back home. The aim of investigating these two periods was to understand their decision to return to Albania and the impact of their migration experiences on their integration back into modern Albanian society.

To analyse the interviews we used the method of thematic analysis proposed by Braun and Clarke.[16] This method consists of six steps: familiarization with the data, coding, searching for themes, reviewing themes, defining and naming themes, and writing up. It provided a rich thematic description of the whole data set, as well as allowing us to move back and forth through the steps when exploring new or under-researched areas, as was the case with the reintegration of Albanian returned migrants.

Motivations for return

There have been numerous attempts to draw up a typology on return migration, to offer an in-depth explanation of the reasons that motivate migrants to return home.[17] Some portray return migration as a failure of the migrants to adapt to the economic and socio-cultural norms of the receiving country, thus not fulfilling their initial goals and, hence, pushing them to return home.[18] Others regard it as a successful achievement of migrants' objectives in the receiving country, facilitating their return home.[19] However, it has been pointed out that return migration cannot be explained only by one factor, but rather by a combination of different conditions that complicate any attempt to establish a typology of return motives.[20] The decision to return is thus an often complex and confusing issue for migrants. In cases of couples it was made jointly, although we found it was proposed primarily by the husband. There was only one exception (see the case of Jorgo below), where the decision was made purely by the husband. Among the other cases, only three migrants, one man and two women, were initially reluctant to return. However, after lengthy discussions with their partner, they became convinced that it was the right decision to make at that moment. Klara, 35 years old, told us:

> My husband wanted to set up our business [printing on canvas] in Albania. When we discussed the idea of returning, I was afraid. Afraid of feeling like a foreigner, as I was [feeling] in the beginning in Greece. [Afraid] [t]o start over again. I didn't want to go through it again. But one summer we went to Durrës [Albania] for holidays ... I enjoyed it very much. ... [W]hen we returned to Greece, my husband reopened the discussion of returning and I said 'Yes'.

In the narratives of the returnees we uncovered a variety of reasons that emerged as decisive motivational drives to return, a mixture of push and pull factors, as a response to the changing circumstances in Greece as well as in Albania. Most of the returnees referred to the financial crisis that hit the Greek economy in 2009 as the main determinant, either direct or indirect, of their decision to return. Being unemployed for a considerable period of time in Greece, or seeing their returns diminish significantly, and thus being compelled to use their savings to get by, resulted in the only option left, the decision to return home. Tom, 41 years old, told us the following:

> I returned because I had no work anymore. My wife did not work there [in Greece] because I was working in construction and was earning a lot of money, especially during the Olympic Games in 2004. No way would I return to Albania! But we were spending our savings and we realized that this was not a solution. So we packed all our things and came here. At least here I don't pay rent. I own my house.

Jorgo, 45 years old, returned for the same reason, although his decision was not well-received by his family:

> One day I asked myself 'What am I doing here?' I had no work, no income, and my employer owed me a lot of money. ... So one day I told my wife to pack everything. We were leaving Greece. She and my daughter were surprised and tried to make me change my mind, but I was determined to return to Albania.

Other push factors, in addition to the looming debt crisis but closely related to it, were unanticipated developments such as business failures or the long process of family reunification. Nardi, 35 years old, failed as a businessman after the bar he and his partners opened in Athens went bankrupt. This failure led him into a depressed state that did not get better with time.

> I was very upset because my business did not go well. I was bankrupt, my partners destroyed me. I lost everything and I began to drink. I felt lost in Athens, and there was no reason for me to stay there anymore.

Since the beginning of migration in the early 1990s, the vast majority of Albanian migrants in Greece have had serious problems with their documents. The legalization process that started in 1998 turned out to be long and costly, especially for those who tried to bring their families to Greece. Family reunification was discouraged by the bureaucratic procedures of the Greek government.[21] Hopeless and demoralized as they were, the decision to return came unexpectedly for many of these migrants. As Ben, 35 years old, explains:

> I was tired of Greek bureaucracy. I was waiting too long to bring my wife to Athens, but nothing concrete was being done. It was a matter of time and money. So many years alone! 'What am I doing here?' I asked myself.

Changing economic conditions at home functioned as pull factors for returning.[22] Economic models of migration conceptualize migrants' behaviour as a response to greater possible earnings elsewhere. Jani, 45 years old, returned to Albania because his earnings were higher there than in Greece. The increase in taxes that is usually associated with an economic crisis, which was precisely how the Greek government proceeded to tackle its deficit, had a negative impact on Jani's profits.

> I came to Albania because my earnings are higher. Here I have a coffee shop that is going okay. In Greece I had a restaurant, but because of the high taxes imposed by the Greek government during the economic crisis, it did not go well.

The good economic prospects of the Albanian economy, as shown by its positive, though fluctuating, GDP since 1998,[23] was perceived by the migrants as a better environment (always in relation to their prospects in Greece) for finding a good job, starting a career, or setting up a small business, without necessarily increasing their earnings in relative or absolute terms. Those who had studied and obtained degrees, either at Greek or Albanian universities, were eager to seek work opportunities in Albania related to their fields of study or experience. Their accounts demonstrated a need for professional advancement. Blerina, 43 years old, told us:

> In Albania I was a teacher. ... There [in Greece] I worked as a cleaner in different houses. Although the Greek ladies treated me well, I did not like that kind of job. Very often I asked myself 'What am I doing here?' I knew I was able to do better jobs. ... So one day we decided to return. Now I am a teacher again at a private school, and my husband is an engineer.

The need for professional advancement and the chance to reactivate one's career is understood if we take into account that this group of educated migrants was constrained to work in low-status jobs that Greek workers shunned, such as cleaning, construction, domestic nursing, etc. Artan, 42 years old, confirmed this:

> In Greece I worked for many years in construction, whereas in Albania I was a doctor. We left Albania with the intention to return, after we had accumulated enough money to buy a house. I tried to practise my profession in Greece but [there was] no way. I had to work hard ... and there was no time left to seek information on how to recognize my degree. I was tired of doing that kind of work, not only physically but also psychologically. There I had to deal with people [co-workers] who did not understand me ... My wife also did not feel good cleaning houses. We returned to Tirana and bought a house. Now I am working as a doctor, my wife as a teacher.

The words of Artan also provide one of the reasons why Albanians agreed to work in fields not related to their expertise. In cases where the goal was not to settle permanently in Greece, but to return to Albania after having accumulated some savings, or to find a way to migrate to the US, or another more developed Western country, the migrants could not afford to be selective, at least initially, with regard to work opportunities. Fonsi, 38 years old, who used to work as a teacher, told us:

> I came to Greece believing that I would find a way to go to America. I worked [in construction] ... to make the money needed to pay the person who would help me to go there.

Apart from the language criteria associated with certain professions, legalization problems and the negative image of the Albanian migrant in Greek society were the primary reasons why Albanians were unable to find better jobs.[24] This is stressed by Tina, 42 years old.

> In the beginning we had to cope with the idea that being an Albanian within the Greek society ... [meant being] illegal and criminal. To rent a house or to find a job we had to hide our nationality. ... [W]e were constrained to accept low-status jobs.

The economic incentive that functioned as a pull factor for returned migrants was not the general economic outlook in Albania but, specifically, the opportunities in Tirana. Returning to Albania means settling in Tirana, regardless of where they lived before migration. On 1 January 2014, 27.2% of the Albanian population lived in Tirana,[25] while in 2013 the city accounted for 37.6% of Albanian GDP—by far the largest contribution for any city. Per capita income in Tirana was the highest in Albania, surpassing the country's average by 28%.[26] In 2010, the number of small, medium and large businesses in the city was 39,656, while four years later that number increased by around 25%, climbing to 49,467. By 2014, 41% of businesses in Albania were based in Tirana.[27] In addition to employment opportunities associated with economic development, Tirana was regarded by the returned migrants as a metropolitan city, the only one in Albania, which can provide a similar lifestyle to Greece, as well as providing better education for children. Alket, 41 years old from Gjirokastër, said:

> We lived in Athens and I could not convince my children to live in Gjirokastër, a small town with no future prospects for them. In Tirana they have more chances to follow their studies than in Gjirokastër. Also my business would go better, because the population is larger.

Challenges of reintegration

Reintegration occurs when the migrants return and try to readjust to their home society. While in the original receiving or host country the process of integration occurs between the host community and the migrants, in the sending country it occurs between the returned migrants and those who never left their home.[28] Thus, reintegration means reinclusion, readjustment or reincorporation of the migrants back into their country of origin. In our case, when Albanians migrated to Greece, they had to adapt to the Greek way of living at the expense of their previous attitudes and beliefs. This is especially so because Greece is a long way from being a multicultural society with a widespread tolerance of diversity amongst immigrant communities. When they returned to Albania they had to adapt to the new way of living in Albania, because during their absence significant changes had occurred in Albanian society. For some reintegration can take years, while for others it can never be achieved, thus pushing them to migrate again.[29]

Home has changed

Returning home did not necessarily mean that the migrants returned to their life before migrating. Although our interviewees had meantime visited Albania on a regular basis, usually for vacations during summer, Easter or end-of-year holidays, upon return they somehow felt that they were both natives and foreigners. The first negative impressions came from the changes in the urban fabric of the capital. The city of Tirana had grown chaotically. As Luiza, 43 years old, put it: 'I was born and raised in this city. I left when I was 23 years old. Now I see this city and barely recognise it. Many high buildings without plan, without order.' Jani, 43 years old, told us: 'in the beginning we were foreigners in a foreign country [Greece]. Now we are foreigners in our own country. I know better the streets of Athens than these of Tirana'.

Poor infrastructure has always been a problem in Albania. During communism, the two main inconveniences were frequent power cuts and shortages in the supply of running water. Other relevant problems included a poor urban and interurban transportation network. The situation with electricity and the water supply improved gradually after the fall of communism, but other problems related to uncontrolled and unplanned urbanization emerged, the most serious one being pollution. Internal mobility and external migration led to a high demand for housing in Tirana, which construction companies eagerly supplied by exploiting every vacant space within and outside the city. It was this chaotic urban expansion, with all of its accompanying problems, that return migrants initially noticed. The first year of return was the hardest and the most disappointing, especially for those who returned before 2008. Problems remained with power and water supplies, and, in addition, there were unpaved and dirty streets and squares, as well as many fewer parks and open areas. Flora, 38 years old, returned to Albania in 2000, but she could not endure these problems for even a year. She left for Greece and returned again, seven years later, to Albania. In 2007, the situation had improved, but some problems were persistent and continue to create difficulties.

> When I returned [2000] there was no drinking water. Waste water was being spilled on the streets. Because of the bad smell you should hold your nose. The building where I lived in was very dirty. No one went outside their home to clean. I was surprised. During communism we kept everything so clean. ... Now [2014] it has changed. ... [but] the air is very polluted ... wastes are being burned in every neighbourhood of Tirana. It is not safe to raise a child here.

Almost all of the interviewees engaged themselves naturally in a process of comparing the new Albanian society they found upon return to the old one prior to their migration, and to Greek society. The comparisons were drawn when they wanted to express disapproval and criticism of the changes that had occurred during their absence, either with regard to the city, or to people's behaviour. The fall of communism in Albania was followed by a partial and asymmetric modernization, which not only involved structural changes, but also changes in the relationships between social structures and social agents. One such relevant change was the loss of traditional support networks. In Albania's particular brand of neoliberal modernity, individuals have to rely on themselves, and put themselves at the centre of their own life plans. They have to make their own choices without expecting help from others.[30] Besmir, 45 years old, did not ask for help. He knew that life in Albania had changed its rhythms, and that people could not dedicate more time to friends and family, yet he would have preferred a warmer welcome.

> We did everything by ourselves here [in Albania]. No one helped us. But we did not ask for help. It is normal to expect your friends to show interest in you. You change apartment from one neighbourhood to another and you feel strange. Imagine moving from one country to another! And it is logical to expect your friends and relatives to welcome you, because you need somebody to tell you that everything will be okay. But I don't blame them. The rhythm of life is not slow, even here.

Mimoza, 35 years old, expected her friends and relatives to be more supportive and helpful, but, to her disappointment, they were reluctant to provide the attention she needed: 'I expected them to support me at least psychologically, or, ideally, to help me find a job. I believed that they would help me find a job. But nothing. They have become indifferent.'

The material side of life has prevailed and relationships are based on self-interest rather than friendship. Luan, 45 years old, believes that money has become an important value to Albanians: 'People seek to construct relationships on the basis of interest. To them money is the solution to everything.' Albert, 65 years old, expressed the same concern.

> The idea that you have a lot of money makes you see yourself above the others here. … I can't say that it stands for all, but the majority thinks that money is everything. Whereas in Greece it's the opposite. They [the Greeks] respect you for who you really are, not for your wealth.

Feelings of nostalgia for some of the old values of the traditional Albanian culture, such as honour, respect, loyalty, and mutual help, which are devalued in modern Albania, are evident in the narratives of the returned migrants.[31] As Pirro, 50 years old, explained:

> I remember that during the communist regime we supported each other very much. I worked as a farmer in Albania … If a member of our group could not finish his job on time, we helped him. Now no one does that.

The expectations the returned migrants had of their friends and relatives were not met. This was not only because the others had changed. Some of the returned migrants acknowledged that they have changed as well. During their acculturation to Greek life, the migrants experienced changes in their world views or beliefs (internal changes), as well as in their way of living, skills, etc. (external changes).[32] To Artan, 39 years old, learning the Greek language was the only means to make possible his adaptation to the social and cultural norms of Greek society. During this process of adaptation it was natural to experience change.

> I think we have also changed, not only them [friends and relatives who never migrated]. I am not the same person. If I wanted to be accepted by the Greek society, it was a must to learn quickly the Greek language, which would give me access to their norms and ways of life.

Artan chose to maintain his ethnic identity. He even tried to instil the Albanian language and cultural norms into his children, which seems to have been unusual among Albanian migrants:[33] 'I wanted them to have an idea of their parents' country of origin. They needed to know their roots.'

The majority of returned migrants we interviewed maintained their ethnic identity, with some of its cultural characteristics, to which they added some values of Greek society. Their strategy of acculturation was integration; that is, they maintained a balance between their original culture and the new one. But apparently their integration came with internal changes significant enough to differentiate them from 'the others' (the Albanians who never migrated), with regard to attitudes, beliefs, and behaviour. These differences were responsible for hindering reintegration. The role of the dramatic urban changes in Albania was thus secondary or supplementary to this more existential change amongst the returnees. Returned migrants felt alienated from the others and preferred to make friendships with other migrants who had returned from Greece. Lili, 40 years old, confirmed:

> In the beginning, I did not like to go out because I did not know anyone. I grew up in Berat, and then migrated to Greece. Tirana is something new to me. Then two other families that had returned from Greece moved into my neighbourhood. We became friends immediately.

Corruption

According to a report from the United Nations Office on Drugs and Crime (UNODC) and the Albanian Institute for Statistics (INSTAT),[34] one-third of Albanian citizens believe corruption to be on the rise, while 41% perceive it as a stable yet ever-present element in Albanian society. Corruption is present in every institution in Albania, even in education. It is a culture of corruption, very foreign to returned migrants, who complain about the inconveniences that corruption creates.[35] Those who started their own businesses encountered this problem immediately. They intended to act in accordance with the law, pay their taxes on time, pay their bills, etc., although this was not the usual practice in Albania. Instead, Albanian businesses were more prone to paying off public officials rather than their dues to the state.[36] By itself, this was not necessarily a problem (i.e. what other business-owners do does not directly affect the law-abiding returned migrant). But it becomes an obstacle when the public officials expect regular payments from all businesses. In the case of Pandi, 42 years old, public officials invoked bureaucratic procedures and deliberately created problems in order to extort bribes.

> When I opened my shoe store I paid all my taxes. But a month after I started my business, the tax inspectors started coming, and kept doing so ... to check my records. Eventually they invented a violation and fined me. They told me that I could avoid paying the fine by giving them a bribe ... I could not believe my ears! ... I told what happened to a friend of mine. He said that this was normal in Albania, and that I should give them what they asked if I wanted to keep them away.

Corruption in education and health services is also widespread, creating other problems for returnees. The returned migrants believed that Albanian higher education lacked quality and standards, compared to the education provided during communism or in Greece. Viola, 43 years old, currently working as a teacher, told us that money can literally buy degrees in Albania; attendance is not required. She said: 'You are here, and you see things. You cannot find anyone without a degree. It is like manufacturing degrees and wholesaling them. I feel embarrassed when I say that I am a teacher.'

Every discussion about corruption raised the issue of Albania's healthcare system. The reaction of the returned migrants to our question on this topic was predictable: some smiled ironically, others raised their shoulders or shook their heads from side to side in disapproval, or even replied by asking rhetorically 'Don't you know?' In their initial gestures and the words that followed, disappointment was plainly evident. They were convinced that Albania's health services during the communist regime were much better, as they are in Greece. Margarita, 60 years old, said: 'If you don't have money they [doctors, nurses] don't touch you. This did not happen during communism, or in Greece. In Greece ... the doctors show compassion and try to help you.'

With regard to patient satisfaction, the Greek health care system in fact lags far behind European standards.[37] Yet the returned migrants trust Greek doctors much more than the Albanian ones. As Erion, 35 years old, explained, it is not only the bad conditions at public hospitals that frighten him. More concerning is the lack of doctors' qualifications.

> One day I noticed that my daughter's eyes were red and ached. I went immediately to the eye specialist. He prescribed pills, antibiotics. In Greece doctors don't give antibiotics that easily. ... I decided to go to a doctor in Athens. He told me that it was a cat allergy.

Lack of work culture

Returned migrants make a distinction with regard to work culture: Albanians who never migrated, or who did not work for a considerable period of time in another country, do not have a work ethos and they tend to avoid working hard.[38] Hence, a lack of trust is created between an employer and the employee. According to Tasim, 45 years old, a common excuse is the low salary, but this is unacceptable to him. 'I can't understand this kind of mentality. If you agree to do this job, you should do it properly, regardless of your low or high salary.' He continued by stressing that in Greece the lower-status migrant had to work hard to gain the trust of the employer and be accepted by society. In Albania, questions of status and acceptance are irrelevant. However, there is a tendency to associate certain occupations, such as sanitary workers, domestic helpers, and waiters with shame and indignity. They are not preferred or appreciated by Albanians.[39] Low productivity here can be explained by their reluctance to accept these jobs. Another explanation is the poor working conditions and rude behaviour of managers or administrators towards their subordinates.[40]

Future intentions

The returnees are increasingly disillusioned with Albanian society.[41] The manifold challenges that obstruct their reintegration outweigh the benefits of economic development in Albania, which, for most of them, was the principal factor for deciding to return. Their enthusiasm was dampened as they began to understand that social and cultural reality in Albania was only partially experienced during their sporadic visits and, hence, their idea of homeland was distorted. This is because they visited Albania for vacations: a time for socializing with friends and family, for relaxation and spending, not for working and earning money. It should not come as a surprise, then, that although most of the returnees secured their immediate objectives—they found a relatively good job, bought a house, and have friends—many expressed the desire to migrate again. To some it is just a desire, for instance due to their age, or the impracticality of moving yet again. To others migrating again is a real goal. In cases of deportation, or persistent unemployment at home, it is understandable to search for opportunities in other countries. Some studies confirm it;[42] some of our interviewees do as well. Albert, 39 years old, currently unemployed, would prefer to return to Greece, but due to the ongoing debt crisis, his chances of finding a permanent job there are close to zero.

> I have tried to find [in Albania] a job as a teacher, but [there is] no way. Three years trying. You have to bribe someone with power to make it happen. ... I would prefer to return there [Greece], because I know that environment, the language, but under these circumstances [due to the crisis], I can't. I should find a way to leave Albania and never return.

The desire to leave Albania is stronger among returned migrants between 30 and 40 years old, who migrated at a younger age. Although they work in relatively good jobs, and do not face economic troubles, nevertheless the socio-political situation in Albania pushes them toward migrating again. Aleko, 35 years old, was explicit on this issue:

> Things will never improve! How can they? ... It is not a matter of money any more, but a matter of our children's future. I often think of leaving. Even though I have a good job, I am not satisfied. How can you be satisfied? Look around. Everybody stealing, everywhere corruption, the environment so dirty. You smell car fumes every morning. You buy products unchecked, unwarranted and unhealthy. ... I blame myself for returning here.

The role of the Albanian government as a 'facilitator' of reintegration

The Albanian government has developed the so-called 'Strategy on Reintegration of Returned Albanian Citizens 2010–2015', which aims at facilitating the adjustment of returned migrants to the Albanian reality.[43] It is declared to be an institutional framework for return migration and reintegration. Many central institutions, ministries, or departments within ministries are involved in implementing it. In this case, 'implementation' means the setting up of migration counters in different districts, and information points at the borders, for providing information to citizens who wish to migrate or returnees who intend to stay in Albania. Returned migrants can also participate in social programmes designed to aid the needs of the general public, such as free training programmes offered by local employment offices, or apply for monthly economic assistance of ALL8,000 (approximately €55) per family.[44] Those who live or have settled in rural areas can benefit from different subsidy schemes for the production of agricultural products.

Nevertheless, none of our interviewees had knowledge of any effort by the government to reach them. Almost all of them laughed in disbelief when we informed them about the government's intention to provide assistance. Their expectations from the government were, and still are, very low. This is probably the reason why they did not declare to the authorities, Greek or Albanian, their intention to return to Albania (thus adding to the problem of data collection on return migration). From their experiences in Greece they learned that no government would ever assist them in any way. Marko, 40 years old, very disappointed with the Albanian and Greek governments, told us the following:

> I did not declare at the border that I would return to Albania. Why should have I? They [Albanian embassy in Athens] did not help us in Greece ... We were working hard there, and the Albanian embassy was robbing us by requesting money to release a birth certificate, 30 euros only for a stamp. ... We were left alone there, and now here.

Only two of the returned migrants attended free professional training programmes, but never received any benefits with regard to employment opportunities. Fredi, 35 years old, said:

> I attended a training course for fixing TVs and other electric appliances. … I heard about it from a friend of mine. After completing the course I got a certificate but … I could not find a job, neither as a history teacher [his professional qualification], nor as a technician. I have not seen any other benefits from the Albanian government and I have not heard about any strategy to help return migrants. As far as I know, neither have my friends.

The lack of information and guidance about investment opportunities is more evident in the case of returned migrants who invested their savings in Albania. Their decision to set up a small business was primarily based on their work experience in Greece and on informal market research they did by themselves. Leonard, 35 years old, failed twice in his business endeavours.

> I returned here with the idea of starting up my business … I did a little research and thought that this kind of business [glass facades for shops and buildings] would go well here. I started it, but it did not go well. … It was monopolized.[45] Then I opened a karate school. Even here I failed.

The impact of returned migrants' savings on the development of the sending countries is a reasonably well-explored topic. Their savings can be invested in housing, education, health care, and business development, contributing to lowering unemployment rates at the local level.[46] Yet, although the Albanian government continues to declare that it has developed policies for attracting Albanians who live abroad, in practice the observed outcomes have been very small.[47]

One issue in particular that concerns male returnees is the fate of their insurance contributions in Greece. The years they have worked in Greece and contributed to the insurance scheme do not suffice to make them eligible for a pension.[48] These contributions will be lost unless an intergovernmental agreement is negotiated that combines their service days in Greece and in Albania. This is the only issue they hope the Albanian government will resolve. Ben, 47 years old, said the following:

> [W]e do not know what will become of those [working days in Greece]. So far we have not heard about any agreement between the two countries. It is a pity. All those working years [will be] wasted, if our government does not do anything.

Conclusion

The decision to return to Albania is a complex issue for migrants. Our research shows that returning was primarily unplanned. The Greek economic crisis was cited by most returnees as the decisive factor that pushed them to return home. Other related factors were business failures and family issues. As pull factors for returning to Albania, we identified economic incentives and career opportunities. The Albanian economy was performing relatively well, and the chances of starting up a successful business were perceived as high by the migrants. The more educated migrants were eager to return and start or restart their careers.

Returning generally means settling in Tirana. Tirana has a similar lifestyle to Greece, and provides better employment, educational and leisure opportunities than other parts of Albania, which are perceived as stagnant and unviable. Upon returning, however, the migrants felt like strangers. Albania has undergone vast changes, and the returned migrants needed to reintegrate to the new Albanian society. But their reintegration seems to be psychologically and emotionally fraught. The returned migrants believe that the material side of life has prevailed, relations among Albanians are built on the basis of self-interest, and money has become a value in and of itself. Most of the returned migrants feel they have not been provided the support and attention they needed and expected from their friends and relatives. In addition, poor infrastructure, widespread corruption, lack of a work culture among those who never migrated, as well as lack of support from the Albanian government, are issues that hinder their reintegration.

There is a tendency among the returned migrants to differentiate themselves from those who never migrated. They feel alienated from the non-migrants, and tend to make friendships with other returned migrants from Greece. Comparing 'new' Albania to 'old' Albania and to Greece makes them feel unsettled. They are living in between three worlds, struggling to find where they fit best, and reconsidering the wisdom of their decision to return to Albania. Lack of trust in the Albanian government and the belief that things will not improve have convinced those who are younger that migrating again is a solution.

Notes

1. S. Hall. 'Minimal selves', in H. Bhabha (ed.), *Identity—The Real Me: Post Modernism and the Question of Identity*, ICA Documents, London, 1987, p. 44; G. Gmelch, 'Return migration', *Annual Review of Anthropology*, 9(1), 1980, p. 135.
2. R. King, 'Generalizations from the history of return migration', in B. Ghosh (ed.), *Return Migration: Journey of Hope or Despair?*, IOM, Geneva, 2000, pp. 7–55.
3. J.-P. Cassarino, 'Theorizing return migration: the conceptual approach to return migrants revisited', *International Journal on Multicultural Societies*, 6(2), 2004, pp. 253–279.
4. Gmelch, op. cit.; Cassarino, op. cit.; F. P. Cerase, 'Expectations and reality: a case study of return migration from the United States to Southern Italy', *International Migration Review*, 8(2), 1974, pp. 245–62; R. King, 'Return migration: a neglected aspect of population geography', *Area*, 10(3), 1978, pp. 175–182; A. Constant and D. S. Massey, 'Return migration by German guestworkers: neoclassical versus new economic theories', *International Migration*, 40, 2002, pp. 5–38; C. Dustmann and Y. Weiss, 'Return migration: theory and empirical evidence from the UK', *British Journal of Industrial Relations*, 45, 2007, pp. 236–256.
5. R. King, 'Return migration and regional economic development: an overview', in R. King (ed.), *Return Migration and Regional Economic Problems*, Croom Helm, London, 1986, pp. 1–36; S. Ammasari and R. Black, 'Harnessing the potential of migration and return to promote development: applying concepts to West Africa', *Migration Research Series*, 5, IOM (International Organization for Migration), Geneva, 2001.
6. L. Labrianidis and A. Lyberaki, 'Back and forth and in-between: Albanian return migrants from Greece and Italy', *Journal of International Migration and Integration*, 5(1), 2004, p. 77–106; B. Nicholson, 'Migrants as agents of development: Albanian return migrants and microenterprise', in D. Pop (ed.), *New Patterns of Labour Migration in Central and Eastern Europe*, AMM Editura, Cluj Napoca, 2004, pp. 94–110; L. Labrianidis and B. Kazazi, 'Albanian return migrants from Greece and Italy: their impact upon spatial disparities within Albania', *European Urban and Regional Studies*, 13(1), 2006, pp. 59–74.

7. Albanian Institute for Statistics (INSTAT), 'Population and Housing Census in Albania 2011: main results', Tirana, 2012.
8. ACIT, 'Study on the economic impact of the Greek crisis in Albania', Tirana, August 2012 <https://www.usaid.gov/sites/default/files/documents/1863/USAID%20Study%20on%20Greek%20Crisis.pdf> (accessed 8 January 2018).
9. J. Cavounidis, 'Migration and the economic and social landscape of Greece', *South-Eastern Europe Journal of Economics*, 1, 2013, pp. 59–78.
10. F. E. Racher and S. Robinson, 'Are phenomenology and postpositivism strange bedfellows?', *Western Journal of Nursing Research*, 25, 2003, pp. 464–481; T. A. Schwandt, 'Constructivist, interpretivist approaches to human inquiry', in N. K. Denzin and Y. S. Lincoln (eds), *Handbook of Qualitative Research*, Thousand Oaks, CA, Sage, 1994, pp. 118–137.
11. A. Schutz, 'Concept and theory formation in the social sciences', in M. Natanson (ed.), *Philosophy of the Social Sciences: A Reader*, Random House, New York, 1973, pp. 23–44; P. Berger and T. Luckman, *The Social Construction of Reality*, Greek translation by Costas Athanasiou, Nisos, Athens, 2003.
12. One of the interviewees lives in Tirana, but works in Durrës. The distance between the two cities is approximately 35 km, which allows for a quasi-unification of work and housing markets.
13. INSTAT, 'Migration in Albania. Population and Housing Census 2001', Tirana, 2004; INSTAT Census 2011, op. cit. See also INSTAT, Population of Albania, 1 January 2018 <http://www.instat.gov.al/al/temat/treguesit-demografik%C3%AB-dhe-social%C3%AB/popullsia/#tab2> (accessed 26 April 2018).
14. R. King, 'Albania: interrelationships between population, poverty, development, internal and international migration', Méditerranée, 103(3–4), 2004, p. 46; B. Nicholson, 'From migrant to micro-entrepreneur: do-it-yourself development in Albania', *South-East Europe Review*, 4(3), 2001, pp. 39–41.
15. E. Cela, 'Migration and return migration in later life to Albania: the pendulum between subjective wellbeing and place', in Z. Vathi and R. King (eds), *Return Migration and Psychosocial Wellbeing: Discourses, Policy-Making and Outcomes for Migrants and their Families*, Routledge, London, 2017, pp. 203–220; R. King, E. Cela, T. Fokkema and J. Vullnetari, 'The migration and wellbeing of the zero generation: transgenerational care, grandparenting and loneliness amongst Albanian older people', *Population, Space and Place*, 20(8), 2014, pp. 728–738.
16. V. Braun and V. Clarke, 'Using thematic analysis in psychology', *Qualitative Research in Psychology*, 3(2), 2006, pp. 77–101; V. Braun, V. Clarke and N. Rance, 'How to use thematic analysis with interview data', in A. Vossler and N. Moller (eds), *The Counselling & Psychotherapy Research Handbook*, Sage, London, 2014, pp. 183–197; V. Braun and V. Clarke, *Successful Qualitative Research: A Practical Guide for Beginners*, Sage, London, 2013.
17. Gmelch, op. cit.; Cassarino, op. cit.; Cerase, op. cit.; King, 'Return migration', op. cit.; Constant and Massey, op. cit.; Dustmann and Weiss, op. cit.
18. F. Bovenkerk, *The Sociology of Return Migration: A Bibliographic Essay*, Martinus Nijhoff, The Hague, 1974; Constant and Massey, op. cit.; Cassarino op. cit.
19. D. T. Duval, 'Linking return visits and return migration among commonwealth Eastern Caribbean migrants in Toronto', *Global Networks*, 4(1), 2004, pp. 51–67; D. T. Duval. 'Expressions of migrant mobilities among Caribbean migrants in Toronto, Canada', in R. Potter, D. Conway and J. Phillips (eds), *The Experience of Return Migration: Caribbean Perspectives*, Ashgate, London, 2005, pp. 245–263; E. Thomas-Hope, 'Transnational livelihoods and identities in return migration to the Caribbean: the case of skilled returnees to Jamaica', in N. N. Sorensen and K. F. Olwig (eds), *Work and Migration: Life and Livelihoods in a Globalizing World*, Routledge, London, 2002, pp. 187–20; Gmelch, op. cit.
20. King, 'Return migration', op. cit.; R. King and A. Christou, 'Cultural geographies of counter-diasporic migration: perspectives from the study of second-generation "returnees" to Greece', *Population, Space and Place* 16(2), 2010, pp. 103–119.

21. M. Baldwin-Edwards, 'Semi-reluctant hosts: Southern Europe's ambivalent response to immigration', *Studi Emigrazione*, 39(145), 2002, pp. 27–48.
22. Cerase, op. cit.; R, Rogers, 'Return migration in comparative perspective', in D. Kubat (ed.), *The Politics of Return. International Return Migration in Europe*, Center for Migration Studies, New York, 1984, pp. 277–299; A. H. Richmond, 'Explaining return migration', in D. Kubat (ed.), op. cit., pp. 269–275; S. Callea, 'Different forms, reasons and motivations for return migration of persons who voluntarily decide to return to their countries of origin', *International Migration*, 24, 1986, pp. 61–76.
23. Albania, World Bank National Accounts Data. Retrieved on 20 December 2017. https://data.worldbank.org/indicator/NY.GDP.MKTP.KD.ZG?end=2017&locations=AL&start=1984&view=chart
24. On the negative image of Albanians in Greek society see G. Kapllani and N. Mai, '"Greece belongs to the Greeks!" The case of the Greek flag in the hands of an Albanian student', in R. King, N. Mai and S. Schwandner-Sievers (eds), *The New Albanian Migration*, Sussex Academic Press, Brighton, 2005, pp. 153–172.
25. INSTAT, Population of Albania, 2015 <http://www.instat.gov.al/media/2265/pop_1_janar_2015__press_release.pdf> (accessed 25 April 2018).
26. INSTAT, 'Vjetari Statistikor 2010–2014' [Annual statistics], Tirane, 2015, p. 86.
27. Ibid., p. 121.
28. O. O. Arowolo, 'Return migration and the problem of reintegration', *International Migration*, 38(5), 2000, pp. 59–82; M. Tannenbaum, 'Back and forth: immigrants' stories of migration and return', *International Migration*, 45(5), 2007, pp. 147–175; C. G. Howard, 'The returning overseas executive: cultural shock in reverse', *Human Resource Management*, 13(2), 1974, pp. 22–26.
29. Cassarino op. cit.; K. Kerpaci and M. Kuka, 'Feeling as a migrant even in my homeland: living experiences of Albanian return migrants from Greece', in *Proceedings of the 10th International Congress of the Hellenic Geographical Society*, 22–24 October 2014, Thessaloniki, Greece, Special Session 1, 2015, pp. 1153–1161 <http://geolib.geo.auth.gr/index.php/pgc/article/view/10504/10252> (accessed 10 January 2018).
30. U. Beck, *Risk Society: Towards a New Modernity*, Sage, London, 1992; U. Beck, and E. Beck-Gernsheim, *The Normal Chaos of Love*, Polity Press, Cambridge, 1995.
31. Kerpaci and Kuka, op. cit.
32. J. W. Berry, 'Acculturation: a conceptual overview', in M. H. Borstein and L. R. Cote (eds), *Acculturation and Parent–Child Relationships: Measurement and Development*, Lawrence Erlbaum Associates, Mahwah, NJ, 2006, pp. 13–29; J. W. Berry, 'Conceptual approaches to acculturation', in K. M. Chun, P. B. Organista and G. Marín (eds), *Acculturation: Advances in Theory, Measurement, and Applied Research*, American Psychological Association, Washington, DC, 2003, pp. 17–37.
33. Z. Vathi, *Migrating and Settling in a Mobile World: Albanian Migrants and their Children in Europe*, Springer, Dordrecht, 2015; Z. Vathi and V. Duci, 'Making other dreams: the impact of migration on the psychosocial wellbeing of Albanian-origin children and young people upon their families' return to Albania', *Childhood*, 23(1), 2016, p. 59.
34. UNODC and INSTAT, 'Corruption in Albania: Bribery as Experienced by the Population', 2011, p. 34 <https://www.unodc.org/documents/data-and-analysis/statistics/corruption/Albania_corruption_report_2011_web_small.pdf> (accessed 10 January 2018).
35. N. Mai and C. Paladini, 'Flexible circularities: integration, return and socio-economic instability within Albanian migration to Italy', in A. Triandafyllidou (ed.), *Circular Migration Between Europe and its Neighbourhood*, Oxford University Press, Oxford, 2013, pp. 212–237.
36. UNODC, 'Business, corruption and crime in Albania: the impact of bribery and other crime on private enterprise', 2013, p. 29 <https://www.unodc.org/documents/data-and-analysis/statistics/corruption/Albania_Business_Corruption_2013_EN.pdf> (accessed 11 January 2018).

37. OECD, 'Government at a glance, Greece', 2015 <https://www.oecd.org/gov/Greece.pdf> (accessed 10 January 2018).
38. Kerpaci and Kuka, op. cit., pp. 1158–1159.
39. Mai and Paladini, op. cit.
40. Kerpaci and Kuka, op. cit., p. 1159.
41. A similar case of disillusionment is that of Greek second-generation returnees from Germany and the US. Interestingly enough, Greek returnees' complaints against their countrymen are almost the same as those made by Albanian returnees. Corruption, bad behaviour and insensitiveness towards the environment are the main criticisms for both groups of returnees. King and Christou, op. cit., pp. 111–112.
42. T. Kilic, G. Carletto, B. Davis and A. Zezza, 'Investing back home: return migration and business ownership in Albania', ESA Working Paper No. 07–08, 2007 <http://www.fao.org/3/a-ai460e.pdf> (accessed 11 January 2018); E. Germenji and L. Milo, 'Return and labour status at home: evidence from returnees in Albania', *Southeast European and Black Sea Studies*, 9(4), 2009, pp. 497–517.
43. 'Strategy on reintegration of returned Albanian citizens 2010–2015', June 2010, Albania <http://www.esiweb.org/pdf/schengen_whitelist_project_Strategy%20on%20Reintegration%20of%20Returned%20Albanian%20Citizens%202010-2015.pdf> (accessed 11 January 2018).
44. ACIT, op. cit., p. 16.
45. In this context, the term 'monopoly' is used loosely. It means two things: first, that the price of glass is high, because there are few wholesalers who sell at the same price and, hence, there is little room for profit; second, that to secure a profitable project, one needs to have prominent connections or/and offer bribes.
46. King, 'Generalizations from the history of return migration', op. cit.; Ammassari and Black, op. cit.
47. J. Vullnetari, '"Albania: €1" or the story of "big policies, small outcomes": how Albania constructs and engages its diaspora', in M. Collyer (ed.), *Emigration Nations: Policies and Ideologies of Emigrant Engagement*, Palgrave Macmillan, Basingstoke, 2013, pp. 25–49.
48. Workers in Greece need to have at least 4500 service days, or 15 years, of insured work in order to be eligible for reduced pension. The majority of migrants have worked uninsured for most of their time in Greece. The new social security system was introduced by laws 3863/2010 and 3865/2010, and later amended by laws 4024/2011, 4051/2012, 4052/2012, 4336/2015.

Acknowledgements

The authors would like to thank Terri Marshburn for her helpful language suggestions and corrections.

Disclosure statement

No potential conflict of interest was reported by the authors.

Index

Note: **Bold** page numbers refer to tables; *italic* page numbers refer to figures and page numbers followed by "n" denote endnotes.